A2 LEVEL

PHYSICS
FOR CCEA A2 LEVEL

Revision Guide

2nd EDITION

COLOURPOINT EDUCATIONAL

Pat Carson and Roy White

© Roy White, Pat Carson and Colourpoint Creative Limited 2018

ISBN: 978 1 78073 126 1

Second Edition
First Impression

Layout and design: April Sky Design, Newtownards
Printed by: GPS Colour Graphics Ltd, Belfast

All rights reserved. No part of this publication may be reproduced, stored in a retrieval system or transmitted in any form or by any means, electronic, mechanical, photocopying, scanning, recording or otherwise, without the prior written permission of the copyright owners and publisher of this book.

Copyright has been acknowledged to the best of our ability. If there are any inadvertent errors or omissions, we shall be happy to correct them in any future editions.

Colourpoint Educational
An imprint of Colourpoint Creative Ltd
Colourpoint House
Jubilee Business Park
21 Jubilee Road
Newtownards
County Down
Northern Ireland
BT23 4YH

Tel: 028 9182 6339
Fax: 028 9182 1900
E-mail: sales@colourpoint.co.uk
Web site: www.colourpoint.co.uk

The Authors

Roy White has been teaching Physics to A Level for over 30 years in Belfast. He is currently Head of Department and an enthusiastic classroom practitioner. In addition to this text, he has been the author or co-author of three successful books supporting the work of science teachers in Northern Ireland.

Pat Carson has been teaching Physics to A Level for over 30 years in Belfast and Londonderry.

The authors co-wrote Colourpoint's successful A Level textbooks *Physics for CCEA AS Level* and *Physics for CCEA A2 Level*.

Important Note to Students

This guide has been written to help students preparing for the A2 Physics specification from CCEA. While Colourpoint Creative Ltd and the authors have taken every care in its production, we are not able to guarantee that the book is completely error-free. Additionally, while the book has been written to closely match the CCEA specification, it is the responsibility of each candidate to satisfy themselves that they have fully met the requirements of the CCEA specification prior to sitting an exam set by that body. For this reason, and because specifications change with time, we strongly advise every candidate to avail of a qualified teacher and to check the contents of the most recent specification for themselves prior to the exam. Colourpoint Creative Ltd therefore cannot be held responsible for any errors or omissions in this book or any consequences thereof.

Contents

Unit 4 (A2 1): Deformation of Solids, Thermal Physics, Circular Motion, Oscillations and Atomic and Nuclear Physics.................. 4

4.1	Deformation of Solids	4
4.2	Thermal Physics	8
4.3	Uniform Circular Motion	14
4.4	Simple Harmonic Motion	17
4.5	The Nucleus	25
4.6	Nuclear Decay	28
4.7	Nuclear Energy	32
4.8	Nuclear Fission and Fusion	35

Unit 5 (A2 2): Fields and their Applications 40

5.1	Force Fields	40
5.2	Gravitational Fields	41
5.3	Electric Fields	44
5.4	Capacitors	47
5.5	Magnetic Fields	53
5.6	Deflection of Charged Particles in Electric and Magnetic Fields	61
5.7	Particle Accelerators	65
5.8	Fundamental Particles	67

Answers 71

Unit 4 (A2 1):
Deformation of Solids, Thermal Physics, Circular Motion, Oscillations and Atomic and Nuclear Physics

4.1 Deformation of Solids

Students should be able to:
4.1.1 State Hooke's law and use $F = kx$ to solve simple problems;
4.1.2 Demonstrate an understanding of the terms elastic and plastic deformation and elastic limit;
4.1.3 Distinguish between limit of proportionality and elastic limit;
4.1.4 Define stress, strain and the Young modulus;
4.1.5 Perform and describe an experiment to determine the Young modulus;
4.1.6 Use the equation for strain energy, $E = \frac{1}{2}Fx = \frac{1}{2}kx^2$;
4.1.7 Demonstrate an understanding of the importance of the stress, strain and the Young modulus of a material when making design and economic decisions about materials use;

Hooke's Law

At GCSE the 'elastic limit' and the 'limit of proportionality' were treated as meaning the same thing. In fact, this is not so and at A2 level we must make the distinction clear. Hooke's Law states that:

Up to a maximum load, known as the limit of proportionality, the extension of an elastic material is proportional to the applied load.

Hooke's Law may be written as an equation:

$F = kx$ where F = applied load (N)
k = the Hooke's Law constant (N m^{-1})
x = the extension of the specimen under test (m)

The graph on the right illustrates how the load and extension are related for a typical metal wire. From (0,0) up to the limit of proportionality the line is straight. This is the region where the wire obeys Hooke's Law. Beyond the **limit of proportionality**, the line curves. A point is then reached where any further load will cause the wire to become **permanently** stretched. This is the **elastic limit**.

The elastic limit is therefore the maximum load a specimen can experience and still return to its original length when the deforming force is removed. Up to the elastic limit the material is undergoing **elastic deformation**. Beyond the elastic limit the wire reaches a 'yield point'. The internal molecular structure is being permanently changed as crystal planes slide across each other. **A wire stretched beyond its elastic limit is said to be 'plastic', ie it is undergoing plastic deformation.** It may stretch enormously before it finally breaks.

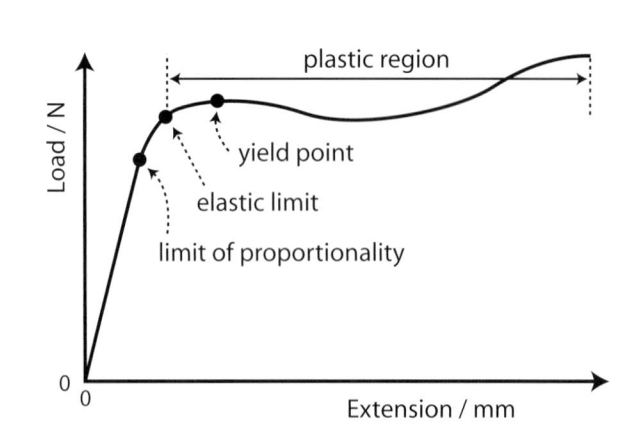

4.1 DEFORMATION OF SOLIDS

Stress, Strain and Young Modulus

The definitions of these quantities need to be learned in preparation for the A2 examination. All of them can be expressed as equations, where:

- σ = stress (Pa)
- ε = strain (no units)
- E = Young Modulus (N m⁻²)
- F is the applied force (N)
- A = cross section area of specimen (m²)
- ΔL = extension of specimen (m)
- L_o = original length of specimen (m)

Stress (σ) is defined as the applied force per unit area of cross section.

$$\sigma = \frac{F}{A}$$

Strain (ε) is defined as the ratio of the change in the length of a specimen to its original length.

$$\varepsilon = \frac{\Delta L}{L_o}$$

Within the limit of proportionality, the ratio of stress to strain is defined as the **Young Modulus** (E).

$$E = \frac{\sigma}{\varepsilon}$$

Measuring the Young Modulus of a Metal

The method below uses two long wires suspended from a common support in the ceiling. One wire is called the reference wire because the extension of the wire under test is measured with respect to it. Both the reference wire and the wire under test should be made of the same material, have the same cross-section area and be approximately the same length.

The wires should be as long as possible (at least 2 m) so as to obtain the greatest possible extension of the test wire.

The length, L, of the test wire is measured in mm using a metre stick.

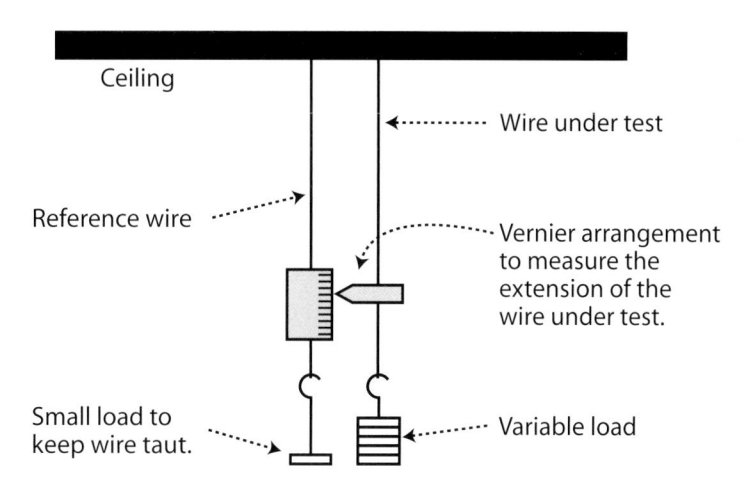

The measurement should be taken from the point of suspension to the Vernier scale. Using a micrometer screw gauge, the diameter, d, of the test wire is measured at about six places spread out along its length. We measure the diameter in this way to avoid the possibility of small kinks in the wire giving rise to erroneous results. The cross section area, A, can then be found from the equation A = ¼π<d>² where <d> is the average diameter of the wire. The reference wire is then loaded with about 5 N to keep it taught. The test wire is loaded in steps of 10 N from 10 N to about 100 N. For each load on the test wire, the extension is found from the Vernier and the stress, σ (σ = F ÷ A) and strain, ε (ε = ΔL ÷ L_o) calculated and recorded in a suitable table.

Typical Results for Young Modulus Experiment on a Metal Wire

This is an experiment prescribed by the A2 1 specification which **candidates must be able to describe**. The table below shows some typical results from the above experiment. The reader should use the results to plot a graph of stress σ (y-axis) against strain ε (x-axis) and draw the straight line of best fit. The gradient of this **straight line** is the Young Modulus.

The gradient of the straight line region (OA) is the Young Modulus.

Length of test wire / m: 2.055
Diameter of test wire / mm: 1.38, 1.38, 1.37, 1.39, 1.38, 1.38
Average diameter of test wire / mm: 1.38
Area of cross section / m²: 1.496×10⁻⁶

F / N	ΔL / mm	σ / MPa	ε ×10⁻⁴
10	0.07	6.68	0.334
20	0.14	13.37	0.668
30	0.21	20.05	1.003
40	0.27	26.74	1.337
50	0.34	33.42	1.671
60	0.41	40.11	2.005
70	0.48	46.79	2.340
80	0.55	53.48	2.674
90	0.62	60.16	3.008
100	0.69	66.84	3.342

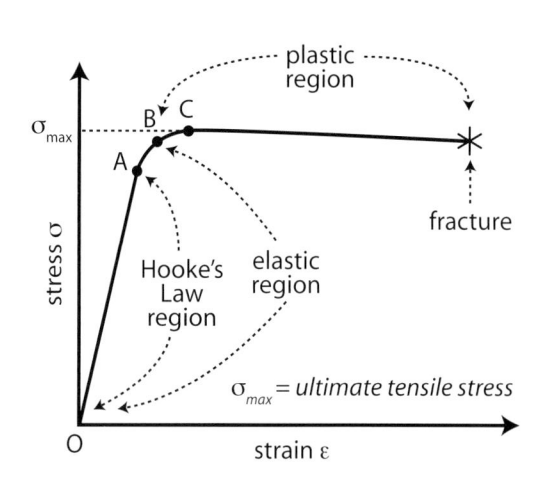

If we continued to measure stress and strain for increasing loads on a wire, we would be able to plot a graph like the one shown to the right of the table. Point A represents the limit of proportionality. Point B represents the elastic limit. Point C marks the position of **ultimate tensile stress** (UTS), defined as the maximum stress which can be applied to a wire without it breaking.

Strain Energy

The graph on the right shows how the extension of a material varies as the stretching force is increased. The work done in stretching the material is stored as potential energy, sometimes known as **elastic strain energy**, E (not to be confused with the Young Modulus, usually also denoted by the letter E). The force varies with the extension, as shown in the graph, and the work done can be determined by calculating the area between the line of the graph and the extension axis.

If the stretching of the material obeys Hooke's Law, and it is not stretched beyond the limit of proportionality, the strain energy can also be calculated as the area under the straight-line graph as shown.

This is the area of a triangle = ½(base × perpendicular height) = ½Fx

Therefore the strain energy $E = ½Fx$

Since Hooke's law is obeyed, $F = kx$
(where k is the Hooke's Law constant measured in N m^{-1}):

Strain energy, $E = ½kx × x = ½kx^2$

Choice of materials

Many factors affect the choice of a material to be used for a particular purpose. These factors include:

- **Stiffness,** which is a measure of how much something stretches when a load is applied, and is represented by the Young modulus. It is a constant of the material and is not affected by the shape or size of an object. Many real-world applications require a stiff material, such as a metal. Examples include car subframes and railway tracks.
- **Lightness**, which is a measure of the density of the material. Some applications require light materials, eg packaging foams.
- **Cost**, another important factor when choosing a material for a particular application. For example, concrete is both cheap and stiff and is ideal for structures such as buildings, while cast iron is ideal for the base of a machine tool bed since it also provides high stiffness at low cost.

Selecting the best material for a particular purpose can be a difficult task and in most cases the use of appropriate computer software is an important part of the decision process. A useful graphical tool is to display two properties of many materials on a scattergram. These scattergrams can be used to compare the ratio of, for example, strength/cost or stiffness/lightness.

In the top scattergram on the right, stiffness is displayed against density and in the bottom diagram stiffness is displayed against cost. Note that the axes are not linear but use a **logarithmic** scale, increasing by factors of ten.

Some applications require materials that are **both** stiff and lightweight, for example rowing oars. Materials with both these properties are not very common but composites offer a good compromise. Their main disadvantage is that they can be quite expensive. Wood is often used for basic rowing oars since it provides stiffness with a relatively low weight. Composite oars are both stronger and lighter than wood, but they are only used by professional rowing teams because they are much more expensive.

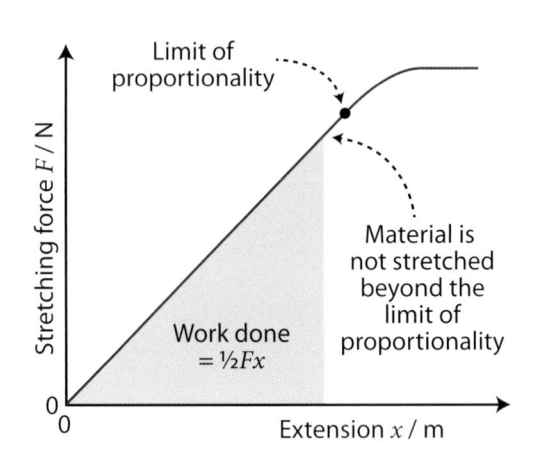

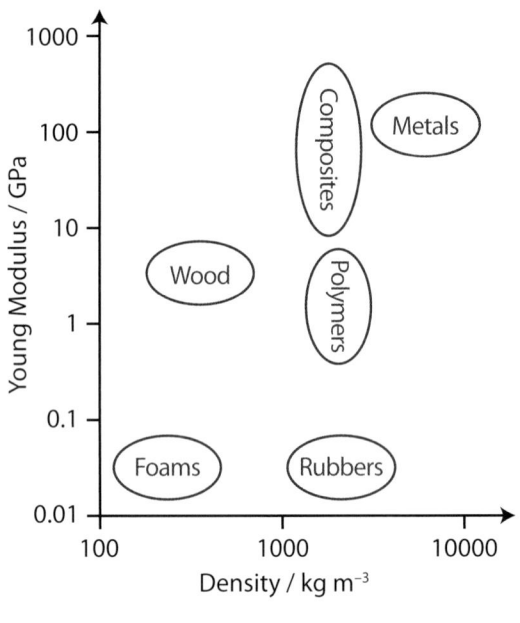

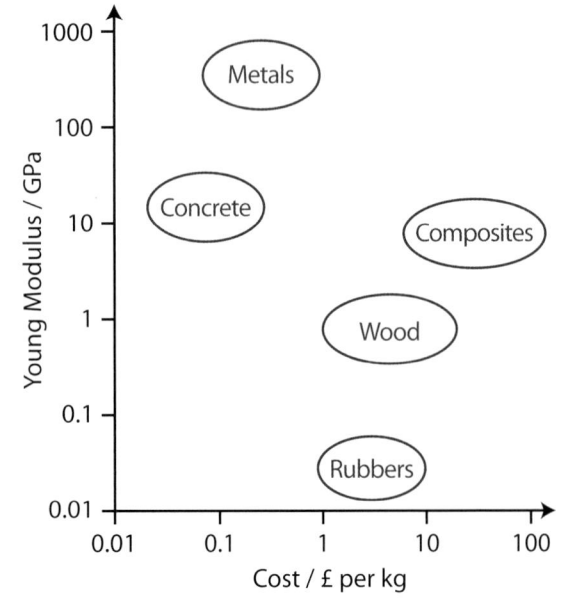

Aluminium frames are used for many applications where a combination of low weight, low cost and high strength is required, for example crutches and zimmer frames, or hang gliders. The covering of a hang glider is usually made from nylon or a similar strong, lightweight material. Boeing's 787 *Dreamliner* will be the first commercial aircraft in which major structural elements are made of composite materials rather than aluminum alloys. The shift is away from fibreglass composites to more advanced carbon laminate and carbon sandwich composites.

Exercise 4.1

1. A metal cube of side 200 mm is held in a vice. Each turn of the handle of the vice moves the jaws 0.500 mm closer together. The vice is tightened up by a quarter turn. A strain gauge attached to the metal shows the compressive force to be 600 kN. Assuming the metal obeys Hooke's Law at this compression, calculate the reduction in the length of the metal and its stiffness constant, k.

2. Two identical springs are joined in series. One has a spring constant of 12 N cm^{-1} and the other has a spring constant of 18 N cm^{-1}. One free end is connected to a fixed point and from the other a weight of 36 N is applied.
   (a) State the tension in each spring.
   (b) Calculate the extension in the combination caused by the 36 N load and the spring (Hooke's Law) constant of the combination.

3. (a) State Hooke's law.
   A spring that obeys Hooke's law has a length of 8 cm when a load of 4 N is attached to it and a length of 8.5 cm when a load of 5 N is attached. Calculate:
   (b) the unextended length of the string.
   (c) the spring constant of the spring and state its units.

4. A copper wire, of length L and cross-sectional area A is stretched by a force F, causing it to increase in length by an amount e.
   (a) Obtain an expression for the Young modulus, E, in terms of L, A, F and e.
   (b) Describe an experiment to determine the Young modulus of steel.

5. A student carried out an experiment to investigate the stretching of a spring as various masses were added to provide the stretching force. The results obtained are shown in the table. below The original length of the spring was 6 cm.

Mass added / kg	Load / N	Length of spring / cm	Extension / cm
0.25		6.8	
0.50		7.7	
0.75		8.6	
1.00		9.4	
1.25		10.2	

   (a) Calculate values for the load and extension and fill in the columns in the table.
   (b) (i) Use values from the table to prove that load is directly proportional to the extension, and hence calculate a value for the spring constant, k.
   (ii) How could a value of k be obtained graphically from these results?

6. A stretching force of 8.0 N is applied to a wire of length 1.5 m, producing a strain of 8.0×10^{-4}. The Young Modulus of the material of the wire is 1.2×10^{11} N m^{-2}.
   (a) Calculate the extension produced in the wire.
   (b) Calculate the cross-sectional area of the wire in mm^2.

PHYSICS FOR CCEA A2 REVISION GUIDE, 2ND EDITION

4.2 Thermal Physics

Students should be able to:

4.2.1 Describe simple experiments on the behaviour of gases to show that $pV = constant$ for a fixed mass of gas at constant temperature, $\frac{p}{T} = constant$ for a fixed mass of gas at constant volume, and $\frac{V}{T} = constant$ for a fixed mass of gas at constant pressure, leading to the equation $\frac{pV}{T} = constant$;

4.2.2 Recall and use the ideal gas equation $pV = nRT$;

4.2.3 Recall and use the ideal gas equation in the form $pV = NkT$;

4.2.4 Demonstrate an understanding of the concept of internal energy as the random distribution of potential and kinetic energy among molecules;

4.2.5 Use the equation $pV = \frac{1}{3}Nm<c^2>$;

4.2.6 Use the equation for average molecular kinetic energy, $\frac{1}{2}m<c^2> = \frac{3}{2}kT$;

4.2.7 Demonstrate an understanding of the concept of absolute zero of temperature;

4.2.8 Perform and describe an electrical method for determining specific heat capacity;

4.2.9 Use the equation $Q = mc\Delta\theta$

The Behaviour of Gases

On a macroscopic scale the behaviour of gases is described by three laws – **Boyle's Law**, **Charles' Law** and the **Pressure Law**. You **need to be able to describe simple experiments** which show these three laws.

Boyle's Law

In this experiment we are investigating the relationship between volume and pressure by keeping the temperature constant. In the diagram on the right, the oil in the closed tube traps a **fixed mass** of air above it. The **length of the air column is proportional to its volume**. The experiment involves measuring the length of this column and recording the corresponding pressure on the Bourdon gauge.

Using the hand pump (or foot pump) we can very slowly increase the pressure acting on the trapped air.

Compressing the gas warms it slightly, so after every compression we need to wait a few moments for the temperature of the trapped air to return to room temperature.

We can repeat this for several more values of pressure and record the new length (volume) and pressure readings in a table.

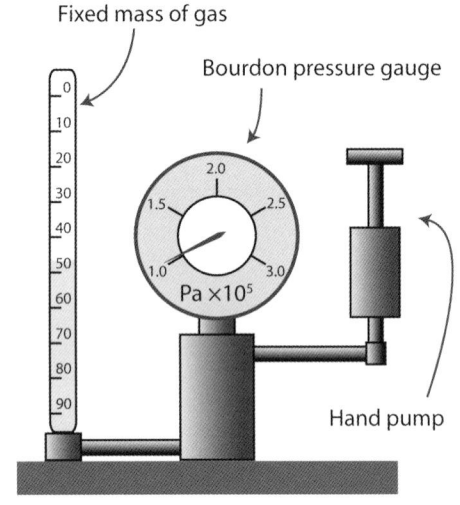

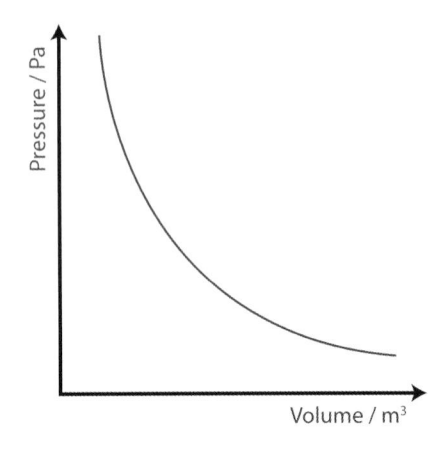

The graph on the left shows pressure plotted against volume. It is a curve which shows that volume decreases as pressure increases.

To determine the nature of this inverse relationship, we plot a graph of volume against 1/pressure as shown in the graph on the right. This graph is a straight line through the origin, confirming **Boyle's Law**:

For a fixed mass of gas at constant temperature, the volume is inversely proportional to the applied pressure.

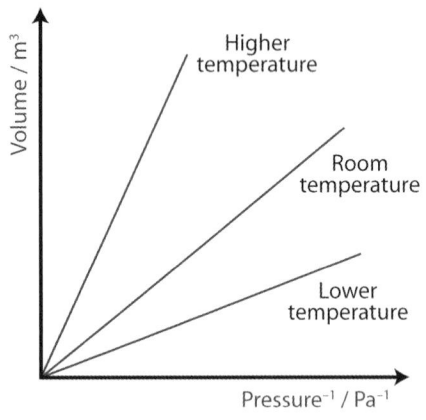

Boyle's Law can be expressed as an equation:

$pV = a\ constant$ where p = the gas pressure (Pa)
V = the gas volume (m³)

Charles' Law

This experiment involves investigating how the volume of a fixed mass of air at constant pressure varies as the temperature changes. As shown in the diagram, air is held inside a glass capillary tube by a short length of concentrated sulphuric acid (the

4.2 THERMAL PHYSICS

concentrated acid traps and dries the air to give better results). The length y of the trapped air is a measure of the volume of the air because the area of cross section of the capillary tube is constant.

The position of the tube is adjusted until the bottom of the trapped air is opposite the zero mark of the ruler. The experiment is carried out at constant pressure: the pressure being exerted on the gas is that of atmospheric pressure and the concentrated sulphuric acid.

The apparatus is then placed in a tall beaker of cold water with a thermometer. Throughout the experiment the water is stirred regularly so that the trapped air is at the same temperature as the water. The volume of the trapped air and the temperature are then recorded in a results table.

The water is then heated until it is about 10°C hotter and another pair of readings of volume and temperature recorded. This process is repeated, increasing the temperature of the water until it boils. A graph is then plotted of volume (on the y-axis) against temperature (x-axis).

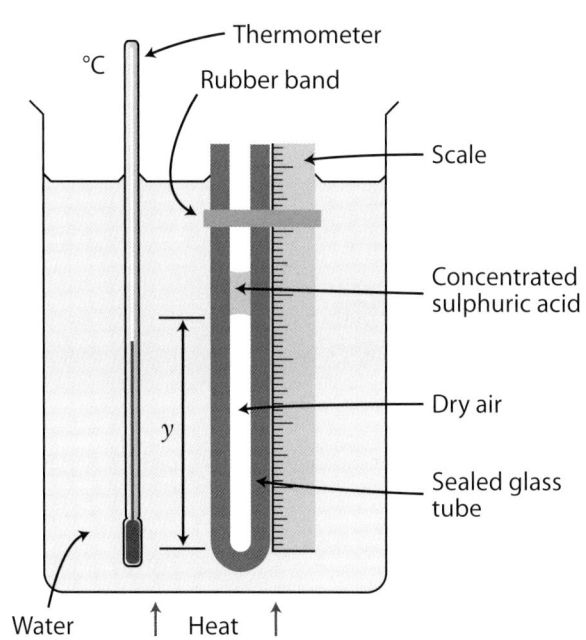

A graph of pressure against the **Celsius** temperature is a straight line of positive slope. However as the graph does not pass through the origin, it does not illustrate proportionality. However, we can use the **Kelvin** temperature scale (K).

The temperature 0 K is called the **absolute zero** of temperature. Absolute zero is the temperature at which **all molecular motion stops** and is approximately −273.16°C, although for purposes of calculations, it is sufficient to use −273°C. There is no temperature below 0 Kelvin. The Kelvin temperature is given by:

$T = \theta + 273$ where T = temperature in Kelvin (K)
θ = temperature in Celsius (°C)

The graph on the right shows pressure against Kelvin temperature. As the Celsius temperature does not pass through the origin, it does not illustrate proportionality. However, we can use the Kelvin temperature scale (K).

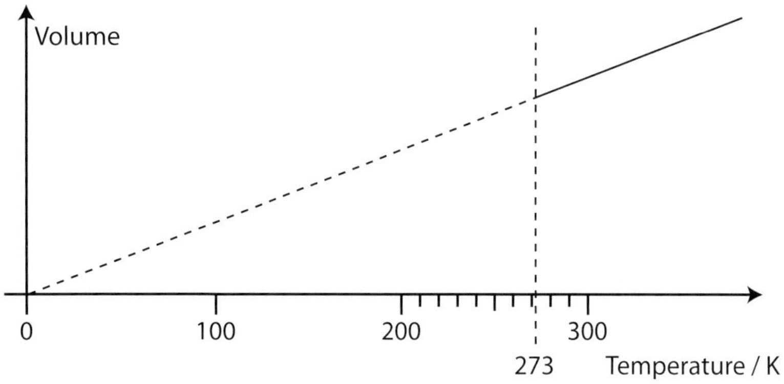

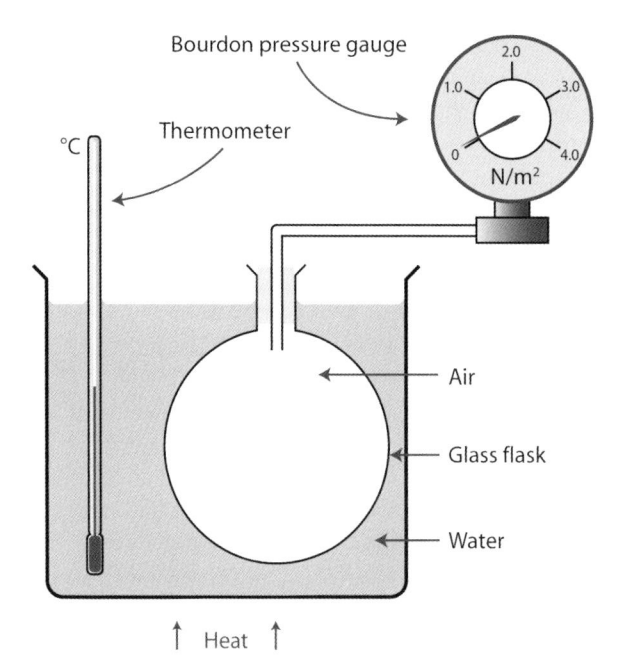

However, a graph of pressure against the **Kelvin** temperature is a straight line passing through the origin, confirming the **Pressure Law**:

For a fixed mass of gas, at constant volume, the pressure (p) is directly proportional to the Kelvin (or absolute) temperature (T).

The Pressure Law can be expressed as an equation:

$\frac{p}{T} = constant$ where p = the gas pressure (Pa)
T = temperature in Kelvin (K)

Ideal Gas Equation

The equations for Boyle's Law, Charles' Law and the Pressure Law can be combined into one, called the **ideal gas equation**:

$\frac{pV}{T} = constant$

If a fixed mass of gas has values p_1, V_1 and T_1, and then some time later has values p_2, V_2 and T_2, then the equation becomes:

$\frac{p_1 V_1}{T_1} = \frac{p_2 V_2}{T_2}$

The Mole

The experimental gas laws show that for a **fixed mass** of gas: $pV = a\ constant \times T$. The constant depends only on the number of molecules in the gas. But the number of molecules present in a given sample of gas is enormous, so a much more convenient way to express this idea is to talk in terms of the number of **moles** of gas in the sample. You will recall that **amount of substance** is measured in moles and that the mole (abbreviated to **mol**) is one of the six SI base units introduced in your AS course. **One mole is the amount of substance which contains as many particles as there are atoms in 0.012 kg of carbon-12.** So, a mole of gas molecules is simply Avogadro's number of those molecules; a mole of electrons is Avogadro's number of electrons and so on. **Avogadro's number is the number of particles per mole**. Its numerical value is $6.02 \times 10^{23}\ mol^{-1}$.

Kinetic Theory and Ideal Gases

The **kinetic theory** attempts to explain the properties of a gas by studying the behaviour of the molecules. In particular, the theory states that it is the collisions of the molecules with the walls of the container that produce an outward force or pressure. These assumptions define the characteristics of what physicists call **an ideal gas**. To apply the kinetic theory we have to make some assumptions. The ideal gas assumptions are:
- There are no intermolecular forces – the only time the molecules exert a force on each other is when they collide.
- The molecules themselves have negligible volume compared to the volume of the gas.
- The collisions between molecules and between molecules and the walls of the container are elastic, so both kinetic energy and momentum are conserved.
- The duration of a collision is negligible compared with the time between collisions.
- Between collisions the molecules move with constant velocity.

Kinetic Theory and the Behaviour of Gases

1. How does the kinetic theory explain the pressure exerted by a gas on the walls of the container?
- Molecules collide elastically with the walls, so each collision results in a momentum change for the molecules.
- A momentum change implies a force was exerted on the molecules by the wall and of course by the molecules on the wall.
- The total force on the wall is the sum of the forces exerted by all the colliding molecules.
- The pressure on the wall is ratio of this total force to the area of the wall.

2. Boyle's Law tells us that when the volume of a fixed mass of gas at constant temperature is doubled the pressure halves. How does the kinetic theory explain this?
- Doubling the volume means that the molecules have on average twice as far to travel to the walls of the container.
- The momentum change per collision is the same.
- So the force caused by each collision is the same.
- But the greater distance means that only half as many collisions occur per second, so the pressure halves.

3. The Pressure Law tells us that as the temperature a fixed mass of gas at constant volume is increased the pressure increases. How does the kinetic theory explain this?
- Increasing temperature increases the speed and momentum of the molecules.
- The momentum change per collision increases and so also does the number of collision per second.
- Both of these contribute to an increase in pressure.

4. Charles' Law tells us that as the temperature of a fixed mass of gas at constant pressure is increased the volume increases. How does the kinetic theory explain this?
- Increasing temperature increases the momentum of the molecules and the collision frequency with the walls.
- To maintain the same pressure the number of collisions per second must decrease.

- Expansion makes the molecules travel a greater distance before they collide with the container.
- A greater distance means a greater time and so the number of collisions per second is reduced, which keeps the pressure constant.

The Universal Gas Constant, R

We can now write down another equation for **an ideal gas:**

$pV = nRT$ where p = the gas pressure (Pa)
V = the gas volume (m³)
n = the number of moles of gas
T = temperature of gas in Kelvin (K)
R = a constant, known as the **universal gas constant**

R has a value of 8.31 J mol⁻¹ K⁻¹ and is a **universal** constant because **it applies to all gases**, provided their behaviour is **ideal**. **This equation is not supplied in the CCEA formula sheet and must be remembered.**

The Boltzmann Constant

The Boltzmann constant, k, is defined by the equation:

$k = \dfrac{R}{N_A}$ where R = the universal gas constant (8.31 J mol⁻¹ K⁻¹)
N_A = Avogadro's number

The Boltzmann constant is therefore 8.31 J mol⁻¹ K⁻¹ ÷ 6.02×10²³ mol⁻¹ and therefore has a value of 1.38×10⁻²³ J K⁻¹. Combining the definitions of the Boltzmann constant and Avogadro's number with the ideal gas equation gives the equation:

$pV = NkT$ where p = the gas pressure (Pa)
V = the gas volume (m³)
N = the number of molecules of gas
k = the Boltzmann constant (1.38×10⁻²³ J K⁻¹)
T = temperature of gas in Kelvin (K)

This equation is important because it links the number of particles in the gas, N, with its macroscopic properties of pressure volume and temperature. **This equation is not supplied in the CCEA formula sheet and must be remembered.**

Linking the Ideal Gas Equation to the Molecular Speeds of Gases

If the speeds of the N molecules in a sample of gas are $c_1, c_2, c_3, \ldots, c_N$, then:

The **mean speed**, $<c>$ is defined by: $<c> = \dfrac{c_1 + c_2 + c_3 + \ldots c_N}{N}$

The **mean square speed**, $<c^2>$ is defined by: $<c^2> = \dfrac{c_1^2 + c_2^2 + c_3^2 + \ldots c_N^2}{N}$

The **root mean square speed**, c_{rms} is defined by: $c_{rms} = \sqrt{<c^2>}$

The relationship between the pressure, volume, mass and speed of the molecules is given by:

It can be shown that the mathematical link between the speeds of the molecules and the gas pressure, p, is given by:

$pV = \dfrac{1}{3}Nm<c^2>$ where N = number of molecules of gas
m = mass of a single molecule (kg)
$<c^2>$ = mean square speed (m² s⁻²)
and the other symbols have their usual meaning

CCEA students do not have to derive this equation. The equation is given in the CCEA formula sheet.

Since $pV = \dfrac{1}{3}Nm<c^2>$ and $pV = NkT$ we can write: $\dfrac{1}{3}Nm<c^2> = NkT$. Multiplying both sides by $\dfrac{3}{2N}$ gives:

$\dfrac{1}{2}m<c^2> = \dfrac{3}{2}kT$ where m = mass of a single molecule (kg)
$<c^2>$ = mean square speed (m² s⁻²)
k = the Boltzmann constant (1.38×10⁻²³ J K⁻¹)
T = temperature of gas in Kelvin (K)

This equation links the average kinetic energy of a collection of gas molecules with the Kelvin temperature.

The Internal Energy of a Gas

The internal energy of a **real** gas is the sum of the potential and kinetic energy of its molecules. However, for **ideal** gases it is assumed that there are **no forces of attraction between the atoms**. So, **ideal gases possess no potential energy**. The internal energy of the molecules of an ideal gas is therefore **entirely kinetic**. The above equation shows that the **average kinetic energy of the molecules in an ideal gas is directly proportional to the Kelvin temperature.**

PHYSICS FOR CCEA A2 REVISION GUIDE, 2ND EDITION

Worked Example

A cylinder has a fixed volume of 1.36×10^{-3} m³ and contains a gas at a pressure of 1.04×10^5 Pa when the temperature is 15°C.
(a) Calculate the number of gas molecules in the container.
(b) Calculate the new pressure of the gas when the temperature is increased to 25°C.
(c) Calculate the increase in kinetic energy of all the gas molecules in the container caused when the temperature is increased to 25°C.

(a) $pV = NkT$ so, $N = \dfrac{pV}{kT} = \dfrac{1.04 \times 10^5 \times 1.36 \times 10^{-3}}{1.38 \times 10^{-23} \times 288} = 3.56 \times 10^{22}$ molecules

(b) From the Pressure Law, $P_2 = \dfrac{P_1 \times T_2}{T_1} = \dfrac{1.04 \times 10^5 \times 298}{288} = 1.08 \times 10^5$ Pa

(c) Average KE of a molecule $= \dfrac{3}{2}kT$, so increase in KE $= \dfrac{3}{2}Nk(T_2 - T_1) = \dfrac{3}{2} \times 3.56 \times 10^{22} \times 1.38 \times 10^{-23} \times (288 - 278) = 7.37$ J

Worked Example

A tyre contains a gas at a pressure of 150 kPa. If the gas has a density of 2.0 kg m⁻³, find the root mean square speed of the molecules.

Since b therefore $P = \dfrac{1}{3}Nm\langle c^2 \rangle \div V$

This means that $P = \dfrac{1}{3} \times \dfrac{\text{mass of gas}}{\text{volume}} \times \langle c^2 \rangle = \dfrac{1}{3} \times \text{density} \times \langle c^2 \rangle$

So $\langle c^2 \rangle = \dfrac{3P}{\text{density}} = \dfrac{3 \times 150 \times 10^3}{2} = 2.25 \times 10^5$

$c_{rms} = \sqrt{\langle c^2 \rangle} = 474$ ms⁻¹

Specific Heat Capacity (SHC)

The specific heat capacity, *c*, of a material is the quantity of heat energy needed to raise the temperature of 1 kg of the material by 1 Kelvin. The units of specific heat capacity are J kg⁻¹ K⁻¹ (or J kg⁻¹ °C⁻¹).

This definition leads to the equation:

$Q = mc\Delta\theta$ where Q = quantity of heat energy supplied (J)
m = mass of the object (kg)
c = the specific heat capacity of the material (J kg⁻¹ K⁻¹)
$\Delta\theta$ = the change in temperature (K)

Experiment to Measure the Specific Heat Capacity of a Metal

The apparatus is set up as shown, with a metal cylinder with two holes in it, one to hold an electrical heater and other to hold a thermometer. A small amount of oil in the hole containing the thermometer is used to improve the thermal contact between the thermometer and the metal.

The mass, m, of the metal cylinder is measured using a balance and the **initial temperature, T_1**, of the metal is measured with a thermometer. The amount of energy can be found using a voltmeter, ammeter and stop clock.

The power of the heater = current × voltage = $I \times V$.

The stop clock is used to determine the time, t, for which the heater was switched on.

Energy supplied, $Q = I \times V \times t$

The final temperature is taken as the **highest temperature, T_2,** reached by the block **after the heater is switched off**. The temperature rise, $\Delta\theta$, is therefore $T_2 - T_1$ and the SHC is calculated as $Q \div m\Delta\theta$. A similar method is used to find the specific heat capacity of a liquid. In this case, the container used to hold the liquid is known as a calorimeter.

The experimental value for SHC is generally larger than the true value because heat is lost from the material under test. To reduce heat loss the metal block or calorimeter is generally wrapped in an insulator, such as expanded polystyrene.

Another technique is to cool the metal to a temperature of around 5°C **less** than room temperature. Heating continues until the metal or liquid temperature is about 5°C **above** room temperature. During the time when the material is below room temperature heat is absorbed from the environment. When the material is above room temperature, heat is lost to the environment. By doing this it is hoped that the heat lost to the environment cancels the heat gained from the environment, and results in a value for the specific heat capacity closer to that which is generally accepted.

The worked example below shows how allowance can also be made for the heat lost to the calorimeter in an experiment to find the SHC of a liquid.

Worked Example

In an experiment, 240 g of milk is heated from room temperature in a container of mass 75.0 g made from copper of specific heat capacity of 390 J kg^{-1} °C^{-1}. A small electrical heater is placed in the milk. The potential difference across the heater is 12.0 V, the current through it is 2.60 A and the heater remains on for 6 minutes and 50 seconds. During this time, the temperature of the container and its contents increases to 13 °C above room temperature. Calculate:
(a) the total heat supplied by the heater,
(b) the heat absorbed by the copper, and
(c) the specific heat capacity of the milk.

(a) Heat = power × time = $(IV) \times t$ = 2.60 × 12 × 410 = 12792 J ≈ 12800 J
(b) $Q = mc\Delta\theta$ = 0.075 × 390 × 13 = 380.25 J ≈ 380 J
(c) Heat absorbed by milk = 12792 − 380.25 = 12411.75 J. SHC, $c = \dfrac{Q}{m\Delta\theta} = \dfrac{12411.75}{0.24 \times 13}$ = 3978.125 ≈ 3980 J kg^{-1} K^{-1}

Exercise 4.2

1. (a) A student gives the following incomplete statement of one of the laws for an ideal gas:

   "The volume of an ideal gas is inversely proportional to the pressure applied to it."

   Identify two important omissions from the correct and complete version of this statement.

   (b) Describe an experiment to investigate the law referred to in (a). Include a labelled diagram. Indicate how you would process your results to clearly demonstrate the relationship between pressure and volume.

2. The air pressure inside a car tyre is 270 kPa at 12°C. After a journey the temperature of the air inside the tyre rises to 25°C. Calculate the air pressure in the tyre. Assume the volume of the air remains constant.

3. A flask contains air at 20°C and is sealed with a rubber bung as shown below. The capillary tube has a diameter of 5.0 mm. A short column of mercury is inserted into the capillary tube as shown. The volume of trapped air is 50 cm^3. The flask is warmed to a temperature of 40°C. Calculate how far up the capillary tube the mercury moves.

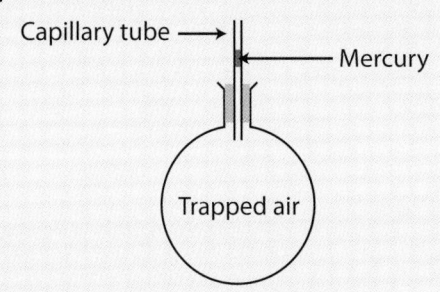

4. When an experiment to demonstrate Boyle's Law was carried out a linear graph of pressure (in Pa), on the y-axis against 1/Volume (in m^{-3}), on the x-axis was obtained, as shown below. Calculating the gradient of the graph gave a value of 1.5×10^4 Pa m^3. The temperature of the gas was 5°C. Calculate the number of moles of gas used in the experiment.

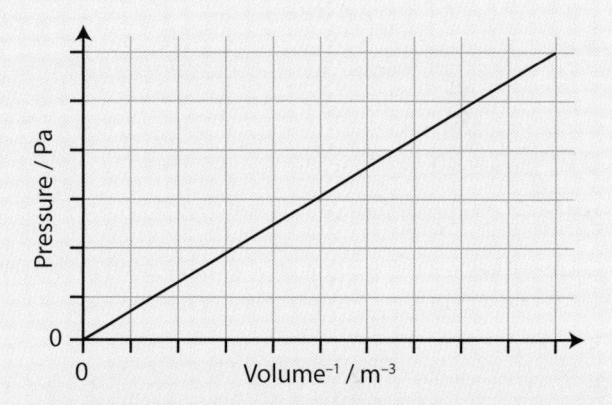

5. The kinetic theory of gases gives the following equation:
   $pV = \frac{1}{3}Nm<c^2>$.

   (a) Explain the meaning of each symbol in the right-hand side of this equation.
   (b) List five assumptions that are made in the derivation of this equation.
   (c) Use this equation to derive a relationship between pressure and density of a gas.

6. (a) Define the terms 'molar mass' and 'Avogadro's number' and explain the link between them.
   (b) Two cylinders of equal volume and at the same temperature each contain 8 kg of gas. One contains molecular hydrogen (relative atomic mass 2), the other contains atomic helium (relative atomic mass 4). Compare the pressures in the two cylinders.

7. The density of air at room temperature is 1.10 kg m^{-3} and atmospheric pressure is 1.01×10^5 Pa. Calculate the RMS speed of the air molecules.

8. (a) State the conditions under which an ideal gas obeys Boyle's Law.
   (b) Draw a labelled diagram of the apparatus you would use to investigate Boyle's Law for air.
   (c) Explain what steps you would take to ensure that the conditions you stated in part (a) applied.

4.3 Uniform Circular Motion

Students should be able to:

4.3.1 Demonstrate an understanding of the concept of angular velocity;

4.3.2 Recall and use the equation $v = r\omega$;

4.3.3 Apply the relationship $F = ma = \dfrac{mv^2}{r}$ to motion in a circle at constant speed;

Angular Velocity

Suppose an object is moving at a steady speed v, in a circle of radius r. Suppose the radius vector sweeps through an angle θ radians, in a time t seconds, as the object moves through arc length s from point P to P'. Then we can define:

Arc length, s, is related to the angle θ by the definition of the radian. This relationship is:

$s = r\theta$ where s = arc length (m)
                               r = radius (m)
                               θ = angle swept by radius vector (rad)

Angular velocity, ω, is defined as the angle swept out by the radius vector in one second. So by definition:

$\omega = \dfrac{\theta}{t}$ where ω = angular velocity (rad s^{-1})
                               t = time (s)

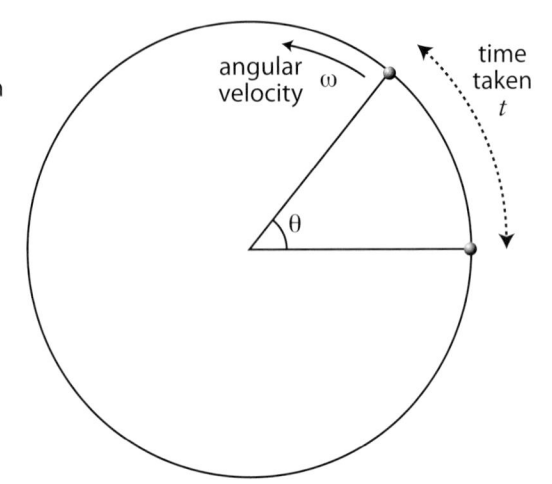

Since **speed, v,** is the rate of change of distance with time, then combining the equations above enables us to write:

$v = \dfrac{s}{t} = \dfrac{r\theta}{t} = r\omega$ where v = speed (m s^{-1})

The **periodic time** of the motion, T, is defined as the time taken for the particle to travel once round the circle through 2π radians. So by definition:

$T = \dfrac{s}{v} = \dfrac{2\pi r}{r\omega} = \dfrac{2\pi}{\omega}$ where T = periodic time (s)

The **frequency, f,** of the motion is the number of revolutions made per second. Since the particle takes T seconds to make one revolution, we can write:

$f = \dfrac{1}{T} = \dfrac{\omega}{2\pi}$ where f = frequency (s^{-1})

Centripetal Acceleration

Any particle moving in a circular path at a constant speed must be accelerating because the direction of its motion, and hence its velocity, is constantly changing. **This acceleration is always directed towards the centre of the circle and is called centripetal acceleration.** The magnitude of the centripetal acceleration is denoted by the symbol a and is given by:

$a = v\omega = r\omega^2 = \dfrac{v^2}{r}$ where a = centripetal acceleration (m s^{-2})

Since the particle is being accelerated there must also be an accelerating force in accordance with Newton's second law. This **centripetal force**, F, is given by the equation:

$F = ma = mv\omega = mr\omega^2 = \dfrac{mv^2}{r}$ where F = centripetal force (N)

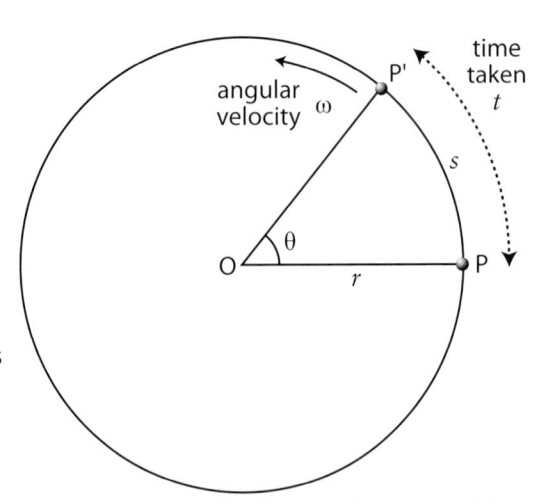

The CCEA specification does not require candidates to derive the equations for centripetal acceleration and centripetal force. However, the ability to recall and use the equations is almost invariably examined.

Causes of the Centripetal Force

Note that the **circular motion does not produce the force**. Rather, **the force is needed for circular motion to take place**. Without this force the object would travel in a straight line along the tangent to the curve. The table on the next page

4.3 UNIFORM CIRCULAR MOTION

identifies the cause of the centripetal force in four different situations.

Physical Situation	Cause of Centripetal Force
A planet orbiting the Sun.	**Gravitational force** between the Sun and the planet.
Electrons orbiting the nucleus of an atom.	**Electrical force** between the positively charged nucleus and the negatively charged electron.
A 'conker' whirled in a circle at the end of a string.	The **tension** in the string.
A racing car going round a circular track.	The **friction force** between the tyres and the track.

Worked Example

The Moon orbits the Earth in a circular orbit of radius 384 400 km. It takes the Moon 27 days to complete one orbit. Calculate:
(a) The angular velocity of the Moon in its orbit, giving your answer in rad s^{-1}.
(b) The linear velocity of the Moon, giving your answer in m s^{-1}.
(c) The centripetal acceleration experienced by the Moon, giving your answer in m s^{-2}.

(a) Using $\omega = \dfrac{2\pi}{T} = \dfrac{2\pi}{27 \times (24 \times 60 \times 60)} = 2.69 \times 10^{-6}$ rad s^{-1}.

(b) $v = \omega r = 2.69 \times 10^{-6} \times 384400 \times 10^3 = 1.03 \times 10^3$ m s^{-1}.

(c) $a = r\omega^2 = 384400 \times 10^3 \times (2.69 \times 10^{-6})^2 = 2.78 \times 10^{-3}$ m s^{-2}.

Motion in a Vertical Circle

The diagram on the right shows an object of mass m being whirled clockwise at a constant speed v in a vertical circle at the end of a piece of string of length L. The resultant force on the object is not constant.

At point A, the tension T_1 in the string is given by:

$T_1 + mg = \dfrac{mv^2}{L}$ so: $T_1 = \dfrac{mv^2}{L} - mg$

At point C, the tension T_2 in the string is given by:

$T_2 - mg = \dfrac{mv^2}{L}$ so: $T_2 = \dfrac{mv^2}{L} + mg$

At points B and D the tension alone provides the centripetal force, so:

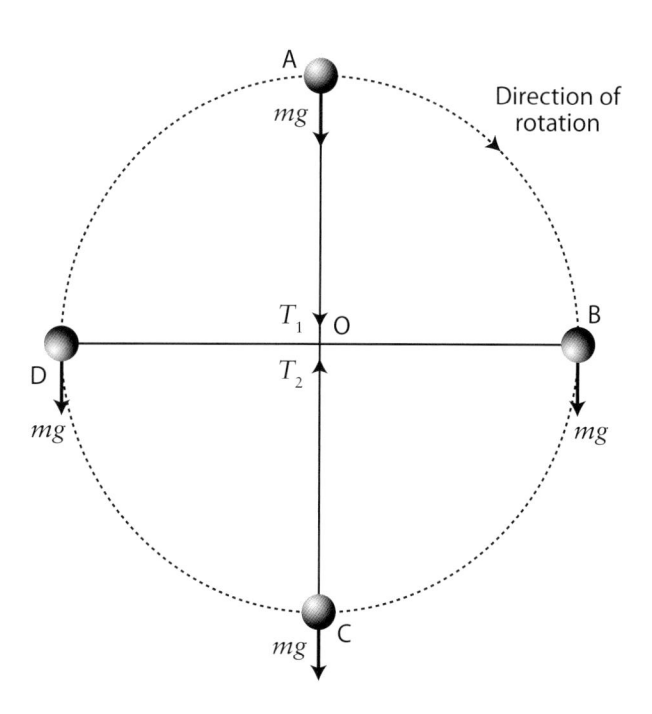

$T = \dfrac{mv^2}{L}$

Circular Motion at an Angle

The diagram on the right shows a conical pendulum. The vertical component tension in the string ($T \sin \theta$), balances the weight of the orbiting mass. The horizontal component of the tension provides the centripetal force. So:

$T \sin \theta = mg$ and $T \cos \theta = \dfrac{mv^2}{R}$

Dividing one equation by the other gives:

$\dfrac{T \sin \theta}{T \cos \theta} = \tan \theta = mg \div \dfrac{mv^2}{R} = \dfrac{Rg}{v^2}$

Note that the object can never be made to move in a horizontal circle, ie where $\theta = 0°$. This would mean that $\tan \theta = 0$. To achieve this, the value of v^2 must be infinite, which is not possible.

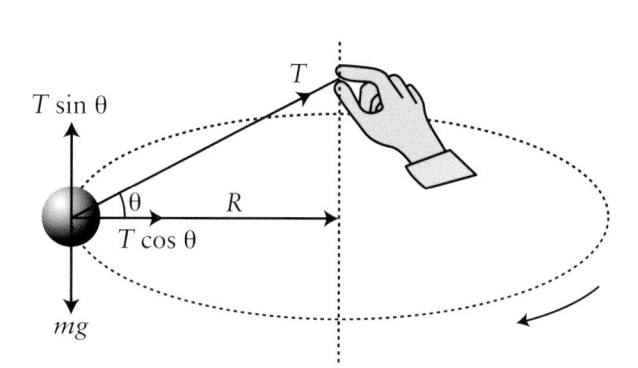

Exercise 4.3

1. An object, attached to a length of string 1.5 m long, is made to move in a horizontal circle of radius 0.4 m as shown in the diagram.
   (a) If the object has a mass of 0.5 kg determine the tension in the string.
   (b) Calculate the value of the centripetal force.
   (c) Calculate the tangential velocity of the object.
   (d) Calculate the period of rotation of the object.

2. A car drives over a hump-back bridge which has a radius of 10.0 m. The speed of the car is 15.0 m s⁻¹. The driver has a mass of 65.0 kg.
   (a) Calculate the reaction force between the driver and the seat when the car is at the top of the bridge.
   (b) Calculate the speed needed if this reaction force is to be zero.

3. An object of mass 0.50 kg is made to move in a vertical circle attached to a length of string. The radius of the circle is 1.0 m.
   (a) Calculate the time needed to complete one orbit if the string is just taut (ie, the tension in the string is almost zero). At what point in the motion will this happen?
   (b) Calculate the maximum tension in the string when it is just taut and state where this occurs.

4. An object of mass 1.5 kg is rotated in a vertical circle on a cord of length 1.2 m. The cord will break when the tension just exceeds 200 N. The speed of rotation is gradually increased from zero. Find the angular velocity at which the string breaks and number of revolutions per second this correspond to.

5. (a) Explain how a body with a constant speed can have centripetal acceleration.
   (b) A student writes down the following:

   "The gravitational pull of the Earth is balanced by the centripetal force acting outwards on the spacecraft and so it stays in orbit."

   Explain why this statement is incorrect and write a corrected version of it.

6. A coin is placed at the edge of a turntable of diameter 25 cm as shown in the diagram. The speed of the turntable is gradually increased.

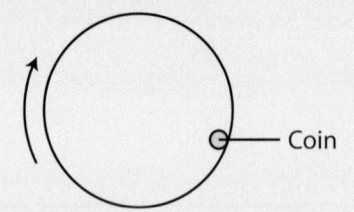

   (a) Explain why the coin will eventually leave the turntable.
   (b) The friction force between the coin and the turntable is equal to 30% of the coin's weight. Calculate the angular velocity of the turntable at which the coin slides off the turntable.

4.4 Simple Harmonic Motion

Students should be able to:

4.4.1 Define simple harmonic motion (SHM) using the equation $a = -\omega^2 x$ where $\omega = 2\pi f$;

4.4.2 Perform calculations using the equation $x = A \cos \omega t$;

4.4.3 Investigate experimentally and graphically the motion of the simple pendulum and the loaded helical spring;

4.4.4 Use the equations $T = 2\pi\sqrt{\dfrac{l}{g}}$ and $T = 2\pi\sqrt{\dfrac{m}{k}}$;

4.4.5 Demonstrate an understanding of SHM graphs, including measuring velocity from the gradient of a displacement-time graph;

4.4.6 Use the terms free vibrations, forced vibrations, resonance and damping in this context;

4.4.7 Demonstrate an understanding of the concepts of light damping, over-damping and critical damping;

4.4.8 Describe mechanical examples of resonance and damping;

Definitions

A particle moves with **simple harmonic motion (SHM)** if its acceleration is proportional to its displacement from a fixed point and the direction of the acceleration is always towards that fixed point. The definition of SHM gives rise to the equation:

$a = -\omega^2 x$ where a = the acceleration (m s^{-2})
ω^2 = a constant (s^{-2})
x = the distance from the fixed point (m)

The minus sign indicates that the acceleration and displacement are in opposite directions. The graphs below illustrate the relationships between acceleration, force and displacement. The amplitude, or maximum displacement, of the motion is denoted by the letter A.

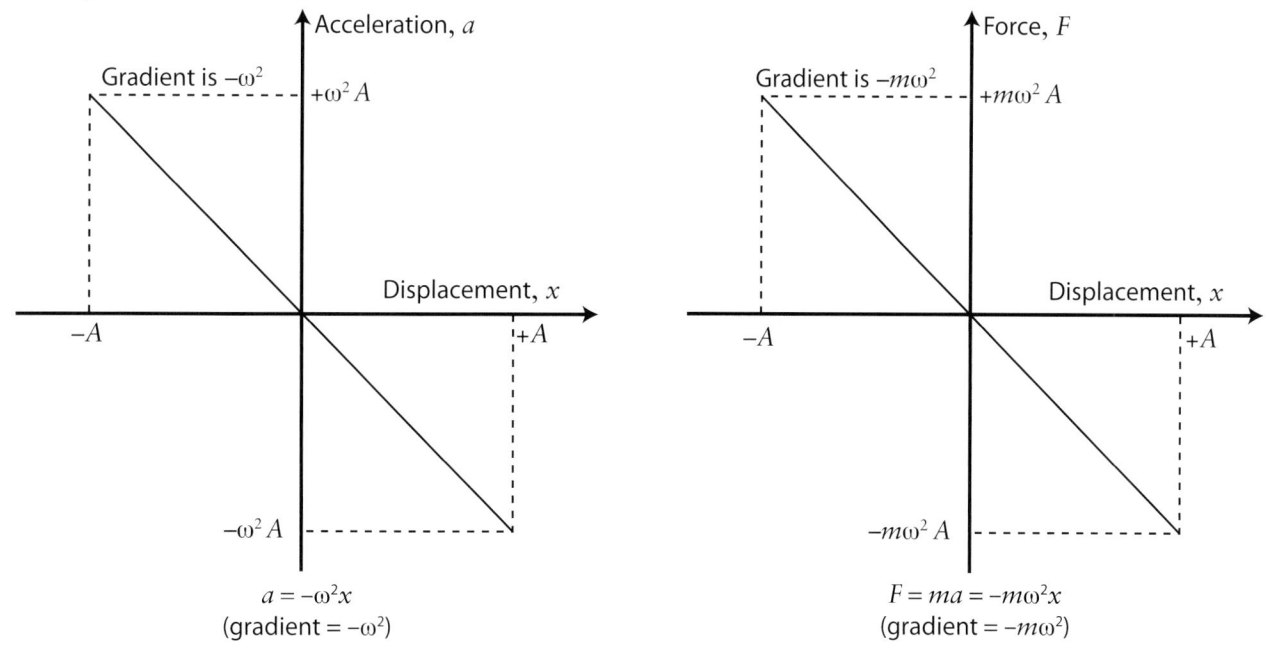

$a = -\omega^2 x$
(gradient = $-\omega^2$)

$F = ma = -m\omega^2 x$
(gradient = $-m\omega^2$)

In simple harmonic motion the object's **displacement** from the equilibrium position, **velocity** and **acceleration all vary with time**. It is not necessary to be able to derive the equations that apply to each of these physical quantities, but it is necessary to be able to use them.

In the diagram below the object R moves along the line YX with simple harmonic motion. The centre of oscillation is the point O. **At time $t = 0$ the object is at Y and moving towards O.**

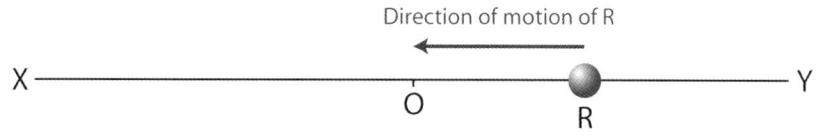

The concept of angular frequency, ω, was first seen when looking at uniform circular motion, and is also applicable to SHM:

$\omega = 2\pi f = \dfrac{2\pi}{T}$ where f = frequency of oscillation (s^{-1})
T = period of oscillation (s)

The following graphs show how the displacement x, (represented by the vector OR), the velocity, v, and the acceleration, a, of the object vary with time, t. The time to complete one oscillation is the period T.

The **displacement**, x, varies sinusoidally, as shown in the graph on the right. The value at any instant is given by the equation:

$x = A \cos \omega t$ where x = displacement (m)
                             A = amplitude (m)
                             ω = angular velocity (rad s^{-1})
                             t = time (s)

Note that given the displacement-time graph, we can obtain the velocity at any instant by drawing the appropriate tangent at the point and finding the gradient.

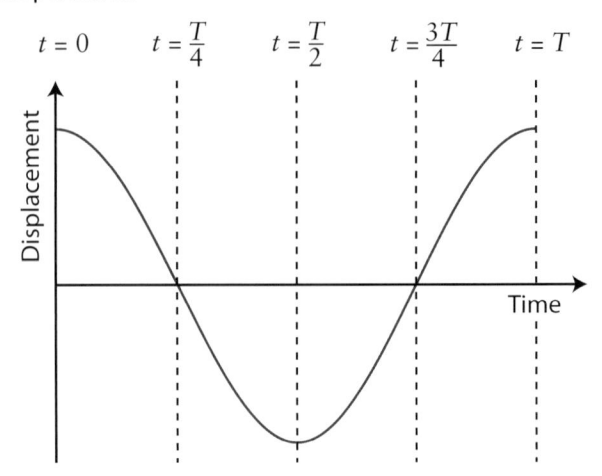

The **velocity**, v, at any instant is equal to the gradient of the displacement-time graph at that instant. The velocity also varies sinusoidally as shown. The general equation for v is:

$v = -\omega A \sin \omega t$ where v = velocity (m s^{-1})
                                     A = amplitude (m)
                                     ω = angular velocity (rad s^{-1})
                                     t = time (s)

Note that the CCEA specification does not require candidates to recall this equation. The velocity has a maximum value of ωA at the instant the object passes through the centre of oscillation.

The minus sign shows that shortly after $t = 0$ the object is moving in the negative direction (from right to left). Note that there is a phase difference between the velocity and the displacement. The velocity is at its peak at time ¾T and the displacement reaches its peak at time T. We therefore say that the velocity is leading the displacement by a quarter of a period, ¼T or ½π radians.

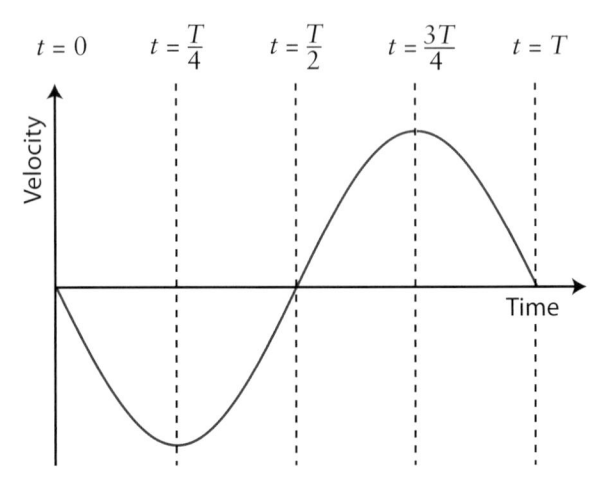

The **acceleration**, a, at any instant is equal to the gradient of the velocity-time graph at that instant. The general equation for a is:

$a = -\omega^2 x = -\omega^2 A \cos \omega t$ where a = acceleration (m s^{-2})
                                                     A = amplitude (m)
                                                   ω = angular velocity (rad s^{-1})
                                                   t = time (s)

The acceleration has a maximum value of $-\omega^2 A$ at the instant the object reaches the extremities of its oscillation. The acceleration is zero when the object reaches the centre of the oscillation.

The minus sign tells us that the acceleration is always in the opposite direction to the displacement from O.

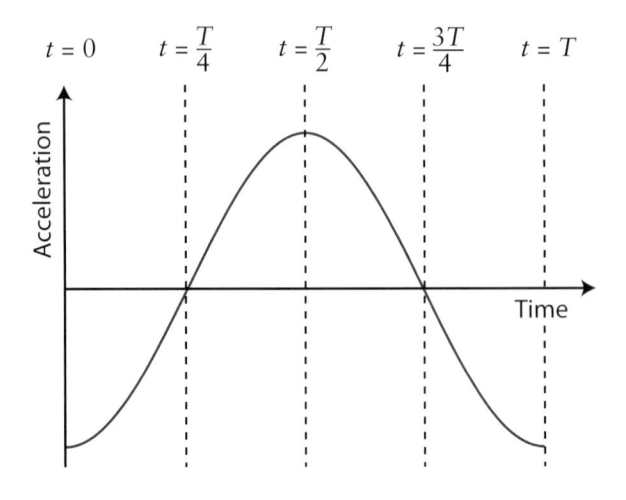

Relationship Between Velocity and Displacement

The defining equation for SHM ($a = -\omega^2 x$) tells us how the acceleration varies with displacement. But how does the velocity vary with displacement? It turns out that:

$v = \pm \omega \sqrt{(A^2 - x^2)}$

The ± indicates that the velocity can be positive or negative, ie to the left or to the right or up or down. In other words it refers to direction. The equation above is **not** required by the CCEA specification, but it is so useful that it is reproduced here for the sake of completeness.

4.4 SIMPLE HARMONIC MOTION

Summary of Equations for Displacement, Velocity and Acceleration

	Displacement	Velocity	Acceleration
Variation with time	$x = A \cos \omega t$	$v = -\omega A \sin \omega t$	$a = -\omega^2 A \cos \omega t$
Maximum value	Amplitude = A	At fixed point, Max velocity = $\pm \omega A$	At extreme displacement, maximum acceleration = $\omega^2 A$
Minimum value	At fixed point, displacement = 0	At extreme displacement, minimum velocity = 0	At fixed point, acceleration = 0
Variation with displacement		$v = \pm \omega \sqrt{(A^2 - x^2)}$	$a = -\omega^2 x$

Worked Example

The graph shows how the displacement of a molecule varies with time in the air as a result of a sound wave passing. The molecule can be assumed to execute simple harmonic motion.
(a) Calculate the frequency of vibration of the molecule.
(b) Calculate the maximum acceleration of the molecule, stating the times at which it occurs.
(c) Explain why the average velocity of the molecule over the 600 µs is zero.
(d) Estimate the average speed of the molecule over the 600 µs.

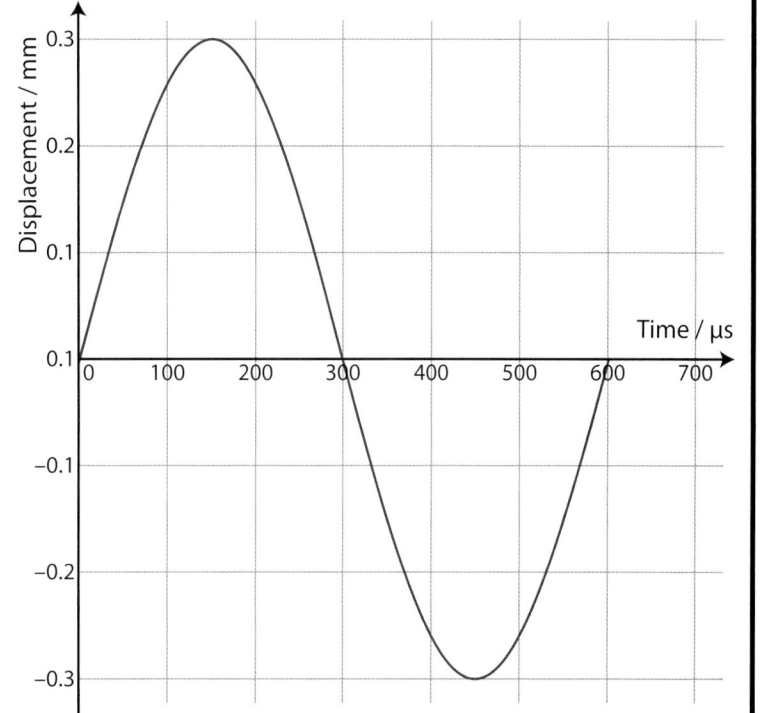

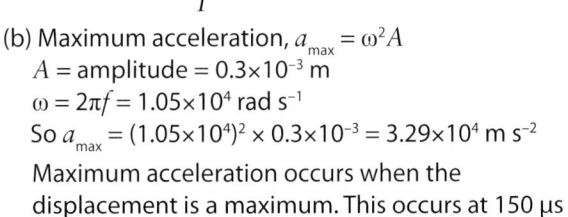

(a) $T = 600$ µs $= 6 \times 10^{-4}$ s
Frequency, $f = \dfrac{1}{T} = 1666.6$ Hz

(b) Maximum acceleration, $a_{max} = \omega^2 A$
A = amplitude = 0.3×10^{-3} m
$\omega = 2\pi f = 1.05 \times 10^4$ rad s^{-1}
So $a_{max} = (1.05 \times 10^4)^2 \times 0.3 \times 10^{-3} = 3.29 \times 10^4$ m s^{-2}
Maximum acceleration occurs when the displacement is a maximum. This occurs at 150 µs and 450 µs.

(c) Average velocity = displacement ÷ time
In one cycle the displacement = 0, therefore average velocity also = 0.

(d) Average speed = distance ÷ time = area under the graph ÷ 6×10^{-4}
We can estimate the area as two triangles: $2 \times (0.5 \times 300 \times 10^{-6} \times 0.3 \times 10^{-3}) = 9.00 \times 10^{-8}$
Thus average speed = $9.00 \times 10^{-8} \div 600 \times 10^{-6} = 1.5 \times 10^{-4}$ m s^{-1}

Worked Example

An object moves in a straight line with simple harmonic motion as shown in the diagram. The centre of the oscillation is marked by the letter O. When the time $t = 0$ the object has a maximum positive displacement. The object undergoes 5 complete oscillations in 10 s.
(a) Calculate the values of A and ω in the equation $x = A \cos \omega t$.
(b) Calculate the maximum acceleration of the object.
(c) Determine the displacement of the object and its position relative to O at a time 0.5 s after the start of the oscillation.

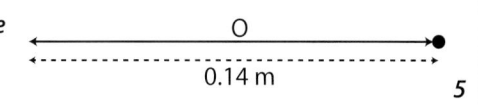

(a) Frequency = $5 \div 10 = 0.5$ s^{-1}
So $\omega = 2\pi f = 2 \times 3.14 \times 0.5 = 3.14$ rad s^{-1}, and amplitude = $A = 0.14 \div 2 = 0.07$ m
The equation therefore becomes $x = 0.07 \cos 3.14t$

(b) Maximum acceleration $a = \omega^2 A = 3.14^2 \times 0.07 = 6.90$ m s^{-2}

(c) $x = 0.07 \cos(3.14 \times 0.5) = 0.07 \cos 1.57$
Note that the value of 1.57 is in radians, not degrees. 1.57 radians = 90°, so we could write this as $x = 0.07 \cos 90°$. Since $\cos 90° = 0$, therefore $x = 0$. So the object is at O.

Motion of a Simple Pendulum and Loaded Spiral Spring

Simple pendulum

The motion of the bob of a simple pendulum is an example of simple harmonic motion. To prove this, let us first look at the forces acting on the bob at the moment it is released from its highest position, as shown in the diagram on the right.

The only forces are the tension, T, in the string and the weight of the pendulum bob $W = mg$. Resolving the weight along the direction of the string gives $mg \cos \theta$. Because there is equilibrium in this direction we can say: $T = mg \cos \theta$

The component of the weight acting perpendicular to the string, F, is given by: $F = mg \sin \theta$

Since no other force acts in this direction, this is the accelerating force.

The displacement from the centre O is x, which is the arc of a circle. Using the equation for angular displacement, $s = r\theta$:

$x = l\theta$ where θ = angle (rad)
                       l = length of the string (m)

This can be rearranged to give: $\theta = \dfrac{x}{l}$

For small angles (10° or under), we can use the approximation: $\sin \theta = \theta$, and so we can write:

$F = mg \sin \theta = mg\theta = \dfrac{mgx}{l}$

Since $F = ma$, we can write: $a = \dfrac{gx}{l}$

As we have seen, the **general** equation for acceleration during SHM, $a = -\omega^2 x$

This can be rearranged to give: $-\omega^2 = \dfrac{a}{x}$

Substituting in the value $a = \dfrac{gx}{l}$ gives us: $-\omega^2 = \dfrac{g}{l}$

The minus sign just indicates that the direction of the acceleration is in the opposite direction to that of the force, so we can omit it and write: $\omega = \sqrt{\dfrac{g}{l}}$

This equation demonstrates that the motion of a simple pendulum meets both criteria for simple harmonic motion:
1. the acceleration is proportional to its displacement, x, from the centre, and
2. the direction of the acceleration is towards the centre point.

Since the general equation for period, $T = \dfrac{2\pi}{\omega}$, for a simple pendulum:

$T = 2\pi \sqrt{\dfrac{l}{g}}$ where T = period (s)
                               l = length of the string (m)
                               g = acceleration due to gravity (m s^{-2})

Loaded spiral spring

When a weight, $W = mg$ is attached to a spring it produces an extension, e, as shown on the right. When it comes to rest at the equilibrium position, O, the tension, T, is equal to the weight, so we can say that $T = mg$. Hooke's Law states:

$T = ke$ where k = the spring constant (N m^{-1})
                      e = extension (m)

Consequently, we can write: $ke = mg$

Suppose that the spring is then pulled down a distance x and released. At the instant of release:

Upward force = $k(e + x) = ke + kx$
Downward force = mg
So the resultant force, $F = ke + kx - mg$
But we know that $ke = mg$, therefore: $F = kx$

The direction is important, because both displacement, x, and force, F, are vectors. Since the force acts in the opposite direction to the direction of the displacement, we add a minus sign to indicate this. Therefore, for a spring:

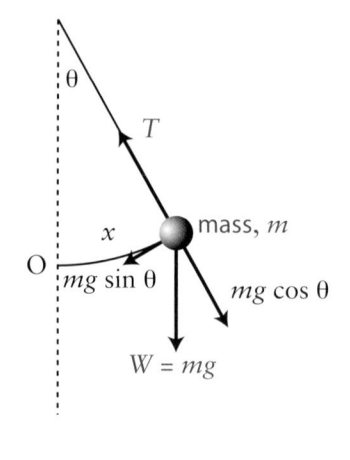

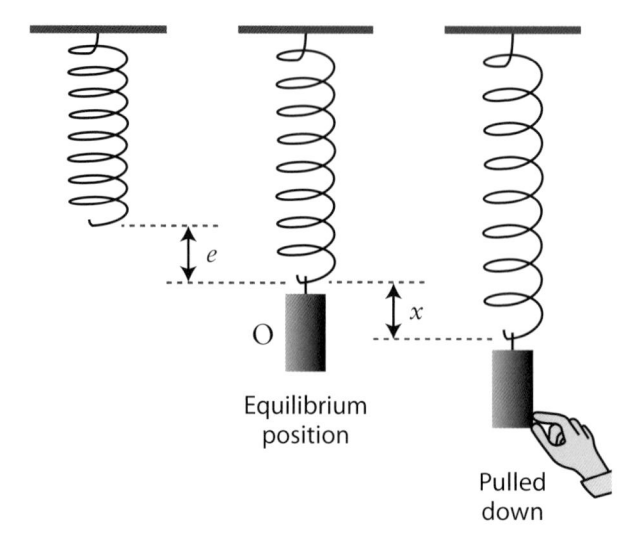

4.4 SIMPLE HARMONIC MOTION

Resultant force at the moment of release, $F = -kx$
Using $F = ma$, the acceleration of the mass at the moment if release is:
$$a = \frac{F}{m} = -\frac{kx}{m}$$
This equation demonstrates that the motion of a loaded spiral spring meets both criteria for simple harmonic motion:
1. the acceleration is proportional to its displacement, x, from the equilibrium position, O and
2. the direction of the acceleration is towards the equilibrium position.

As we have seen, the **general** equation for acceleration during SHM, $a = -\omega^2 x$

This can be rearranged to give: $-\omega^2 = \dfrac{a}{x}$

Substituting in the value $a = -\dfrac{kx}{m}$ gives us: $\omega^2 = \dfrac{k}{m}$

Therefore: $\omega = \sqrt{\dfrac{k}{m}}$

Since the general equation for period, $T = \dfrac{2\pi}{\omega}$, for a loaded spiral spring:

$T = 2\pi\sqrt{\dfrac{m}{k}}$ where T = period (s)
m = mass of the spring (kg)
k = the spring constant (N m^{-1})

Worked Example

A simple pendulum has a length of 0.85 m. The amplitude of oscillation is 4.0 cm.
(a) Calculate its period and frequency of oscillation.
(b) Determine its position relative to the centre of oscillation 1.5 s after it is released from its highest position.

(a) $T = 2\pi\sqrt{\dfrac{l}{g}} = 2 \times 3.14 \times \sqrt{\dfrac{0.85}{9.81}} = 1.85$ s. Frequency $= \dfrac{1}{T} = 0.54$ Hz

(b) $x = A \cos \omega t$, and $\omega = 2\pi f = 3.39$. So $x = 4.0 \times \cos(3.39 \times 1.5) = 4.0 \times \cos(5.085)$.
The 5.085 is in radians, so remember to convert this to degrees. 5.085 rad = 291.5°. So:
$x = 4.0 \cos 291.5° = 4.0 \times 0.3664 = 1.466$ cm for the highest position. So distance from centre = 4.0 − 1.466 = 2.534 cm.

Worked Example

A mass of 0.5 kg causes a steel spring to extend by 8 cm.
(a) Calculate the spring constant.
(b) The mass is allowed to vibrate vertically. Determine the frequency of these oscillations.

(a) $F = ke$, so $0.5 \times 9.81 = k \times 0.08$ giving $k = 61.3$ N m^{-1}

(b) $T = 2\pi\sqrt{\dfrac{m}{k}} = 6.28 \times \sqrt{\dfrac{0.5}{61.3}} = 0.57$ s. Frequency $= \dfrac{1}{T} = 1.75$ Hz

Experimental and graphical investigation of SHM

Simple pendulum

Set up the apparatus as shown in the diagram on the right. The pendulum bob is to be displaced from its equilibrium position – when doing so, ensure the angle between the thread and the vertical is less than 10°. When the bob is released it will oscillate with simple harmonic motion. As shown in the previous section, the periodic time (A to O to B and back to A) depends on the length of the string and the acceleration of free fall.

The length of the pendulum is measured from the just below the suspension point to the centre of the pendulum bob. To determine the period T it is necessary to time at least 10 complete oscillations and calculate an average. It is also possible to use a light gate and computer software to take sufficient and appropriate measurements to determine the periodic time T.

Carry out the experiment for different lengths of thread and tabulate the results in a table like the following:

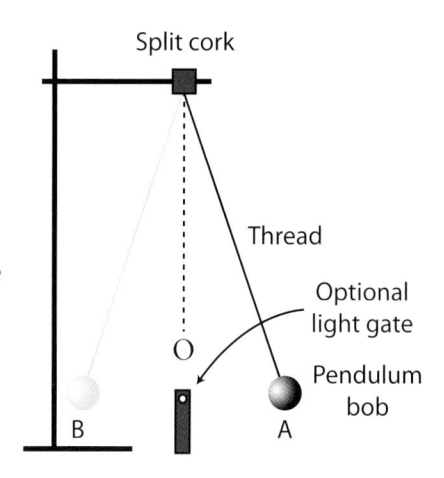

Length of the pendulum, l / m	Time for 10 oscillations / s	Period, T / s
0.2		
0.4		
0.6		
0.8		
1.0		
1.2		

We can now draw a graph from these data. A simple plot of T against l will yield a curve. However we know that the periodic time, T, of the pendulum is given by the equation:

$$T = 2\pi \sqrt{\frac{l}{g}}$$

So, to obtain a **linear** plot, graphs of T against $\sqrt{l}$ or T^2 against l are needed. The gradient of the linear graphs will provide a value for the acceleration of free fall, g.

Loaded spiral spring

Set up the apparatus as shown in the diagram on the right. When the mass is pulled down by a small distance and released the mass will oscillate vertically with simple harmonic motion. The distance the mass is displaced vertically should be small – 5 cm is enough to produce easily-observable oscillations.

The approach to measuring the periodic time is similar to that used for the simple pendulum. To determine the period, T, it is necessary to time at least 10 complete oscillations and take an average. It is also possible to use a light gate and computer software, as shown in the diagram, to take sufficient and appropriate measurements to determine the period.

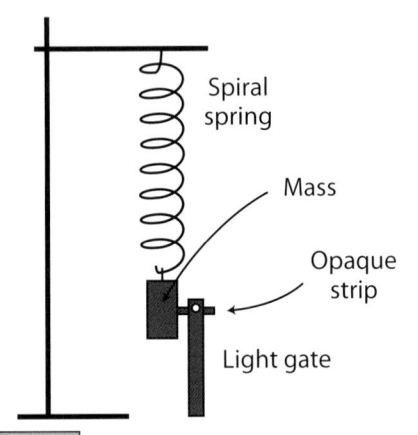

Carry out the experiment for different masses, varying it in steps of 100 g, and tabulate the results in a table like the following:

Mass attached, m / kg	Time for 10 oscillations / s	Period, T / s
0.1		
0.2		
0.3		
0.4		
0.5		
0.6		

We can now draw a graph from these data. A simple plot of T against m will yield a curve. However we know that the periodic time, T, of the pendulum is given by the equation:

$$T = 2\pi \sqrt{\frac{m}{k}}$$

So, to obtain a **linear** plot, graphs of T against $\sqrt{m}$ or T^2 against m are needed. The gradient of the linear graphs will provide a value for the spring constant, k.

Free and Damped Oscillations

If a pendulum is set oscillating the amplitude of the oscillation gradually decreases. Eventually it stops oscillating due to the resistive forces of the air and the friction between the string and the suspension point. The pendulum's energy is gradually transferred to heat and sound. **Resistive forces that act on an oscillating system are known as damping forces. Free oscillations are those which are not damped**, such as the vibration of atoms in a metal.

4.4 SIMPLE HARMONIC MOTION

Undamped oscillations are free oscillations (ie perfect SHM). The displacement varies periodically as shown. However the amplitude remains constant with time.

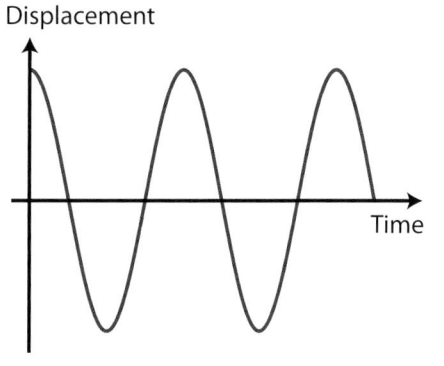

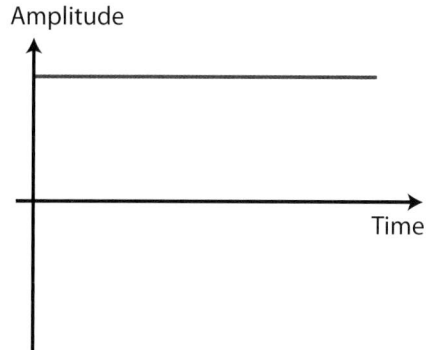

When a system is **lightly damped** the displacement will vary with time as with free oscillations. However the amplitude of the oscillation will gradually get smaller and eventually the oscillations will cease.

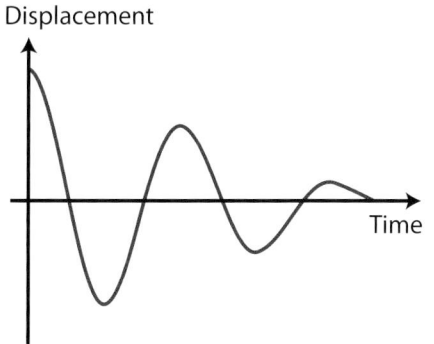

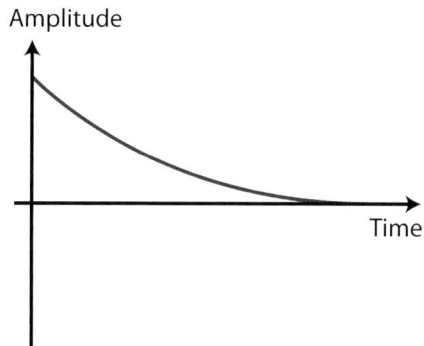

When a system is **over-damped** no oscillations occur. The system returns very slowly to its equilibrium position.

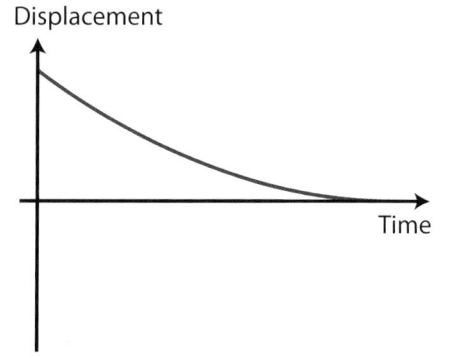

When the damping forces are such that the system can return to its equilibrium position in the shortest possible time then we have **critical damping**.

The shortest possible time for any oscillating system to return to its equilibrium position is ¼T where T is the periodic time.

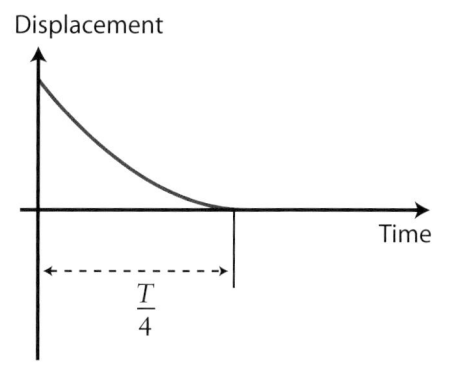

Forced Oscillations and Resonance

A **forced vibration** occurs when any external force which varies with time is used to make an object oscillate. Example include pushing a child on a swing, or making a mass on a spring oscillate by holding it and moving your hand up and down.

Forced oscillations can be demonstrated using **Barton's pendulums** (shown on the right). The 'driver' pendulum X has a heavy bob, while the others are paper cones. The driver pendulum is pulled to one side and released. After a time all the pendulums oscillate with very nearly the same frequency as the driver but with different amplitudes. **The paper cone pendulum, A, which has the same length, L, (and hence the same natural frequency) as the driver, has the largest amplitude.** This is due to **resonance**.

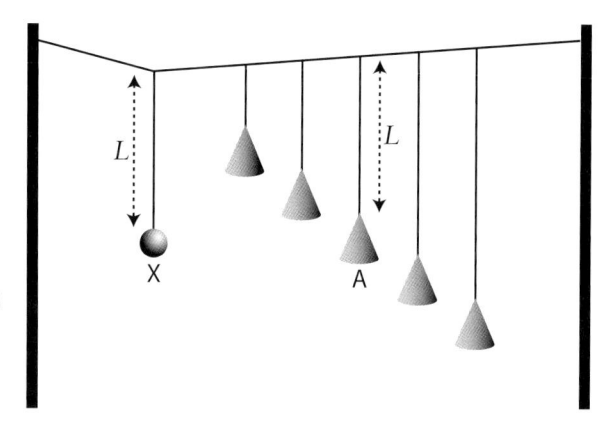

PHYSICS FOR CCEA A2 REVISION GUIDE, 2ND EDITION

Resonance occurs when the frequency of the driving force is the same as the natural frequency of the oscillating system. Any oscillating mechanical system can be made to resonate. Examples of resonance include:
- the air column in a resonance tube
- an oscillating loaded spiral spring
- the vibrating string of a musical instrument
- water molecules in food when placed in a microwave oven.

We recognise resonance when the driven system has maximum amplitude. The graph shows how the amplitude of a forced oscillation depends on the frequency of the force causing it to vibrate (driver frequency). For an undamped or lightly damped system, the amplitude reaches a maximum when the frequency of the driving force equals that of the natural frequency of the oscillating system.

When the damping force is small (light damping) the peak is sharp. However when the damping forces is greater (heavy damping) the **peak is broader** and in fact the maximum occurs **at a slightly lower frequency**. If the damping force is very great the system will not oscillate, so resonance cannot occur.

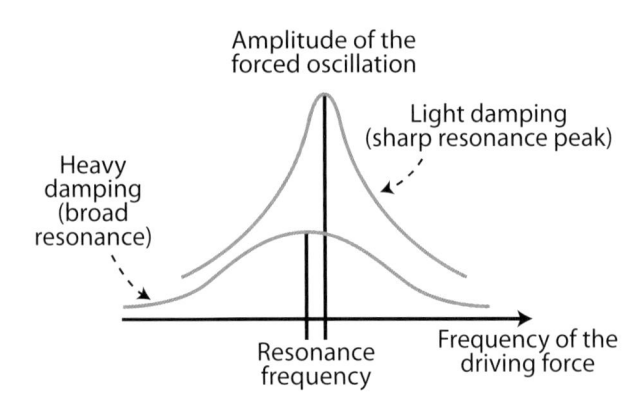

Exercise 4.4

1. (a) Define simple harmonic motion (SHM).
   (b) The acceleration during SHM is given by the equation $a = -\omega^2 x$. Explain the meaning of the symbols used in this equation and the significance of the minus sign.
   (c) Describe how damped oscillations differ from simple harmonic oscillations.

2. The diagram shows a block attached to a spring. When the spring is pulled down a distance of 6 cm and released it undergoes simple harmonic motion with a period of 1.05 s.
   (a) Calculate the maximum acceleration of the block.
   (b) Determine the position of the block relative to its equilibrium position 2.5 s after release.
   (c) Explain why the block will eventually stop oscillating.
   (d) Sketch a graph of displacement (y-axis) and time (x-axis) to show how the displacement of the block varies with time.

3. The equations below show how the period of a simple pendulum and vibrating mass on spring depend on various factors.

   $$T = 2\pi\sqrt{\frac{l}{g}} \qquad T = 2\pi\sqrt{\frac{m}{k}}$$

   (a) Explain the meaning of the terms on the right-hand side of the equations.
   (b) Show that the SI base unit of the left-hand side is consistent with the SI base units of the right hand side.

4. The graph below shows how the tide in a harbour varies. The rise and fall of the tide is an example of simple harmonic motion.

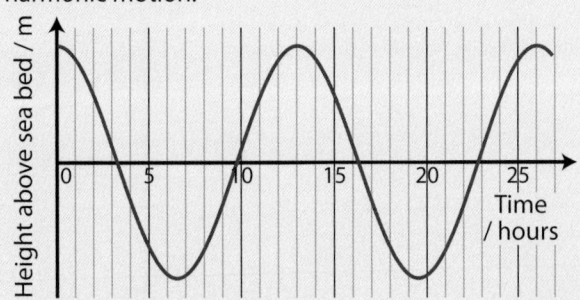

   (a) Using the graph determine the period and frequency of the tides.
   (b) A low tide the water is 1.5 m above the sea bed and at high tide it is 4.8 m above the sea bed. The displacement, X, of the water level is given by the equation $X = A \cos \omega t$. Complete the equation by inserting the values of A and ω.
   (c) By taking the time at low tide as time = 0 how long after this is the water level 2.8 m above the sea bed?

5. Many oscillations are damped.
   (a) What causes damping?
   (b) Sketch a graph to show how:
     (i) the amplitude of a slight damped system changes with time;
     (ii) the displacement of a slightly damped system varies with time.
   (c) What is meant by a forced vibration?
   (d) Under what condition does a forced vibration resulting in a system having a maximum displacement?
   (e) What effect does damping have on the condition described in part (d)?

4.5 The Nucleus

Students should be able to:

4.5.1 Describe alpha-particle scattering as evidence of the existence of atomic nuclei;
4.5.2 Interpret the variation of nuclear radius with nucleon number;
4.5.3 Use the equation $r = r_0 A^{1/3}$ to estimate the density of nuclear matter;

In 1903 Sir J. J. Thomson developed his 'Plum Pudding Model', shown on the right, in which the atom was regarded as a positively charged sphere in which the negatively charged electrons were distributed like currants in a bun in sufficient numbers to make the atom as a whole electrically neutral.

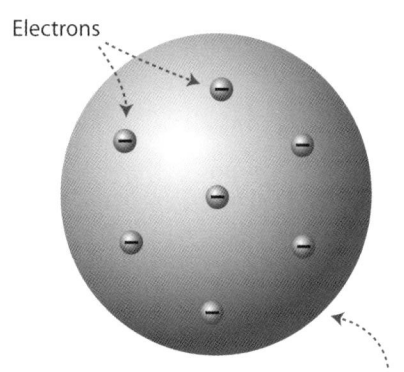

In 1906 Ernest Rutherford noticed that α-particles (positively charged helium nuclei) easily penetrated mica without making holes in it as a bullet might. This led him to suspect that the α-particles were passing right through the atoms themselves rather than pushing atoms out of the way.

Rutherford also noticed that some of the α-particles were deflected out of their straight-line paths as they went through the mica, and he thought that this was caused by electric repulsion between the positively charged part of the mica atoms and the positive α-particles. Rutherford's students, Hans Geiger and Ernest Marsden, then carried out a series of experiments on the scattering of α-particles by thin metal films, as shown in the diagrams below.

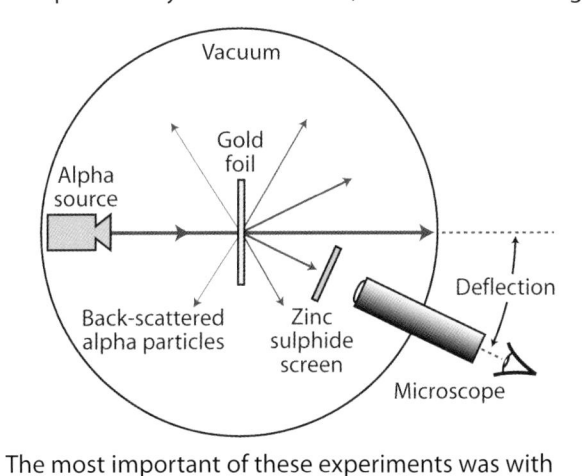

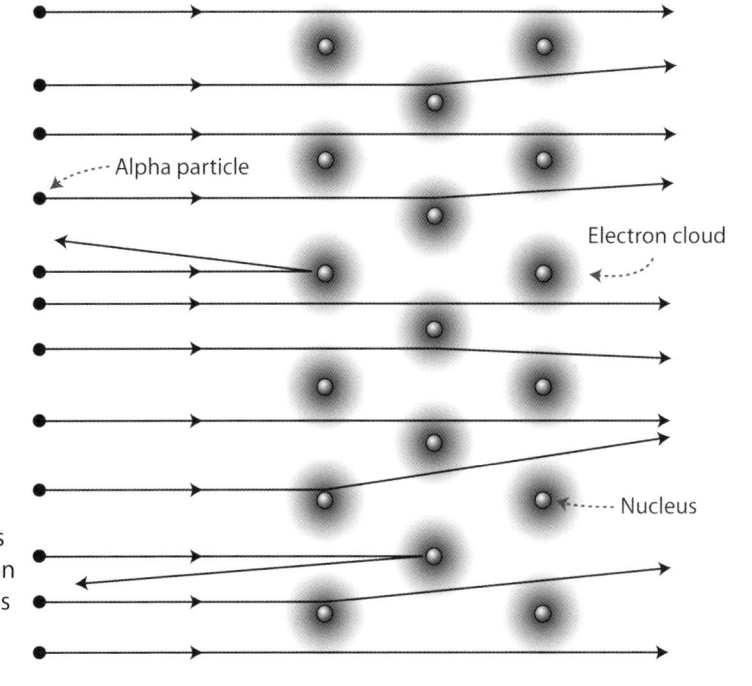

The most important of these experiments was with thin gold foil. A source of α-particles was contained in an evacuated chamber, ie in a vacuum. The α-particles were incident on a thin gold foil (a few hundred atoms thick) whose plane was perpendicular to their direction of motion. The α-particles were detected by the flashes of light (scintillations) they produced when they hit a fluorescent glass screen. The experiment **had to be carried out in a vacuum to prevent collisions between alpha particles and gas atoms** deflecting the alpha particles. The table outlines the results and Rutherford's conclusions.

Results Of Experiment	Rutherford's Conclusion
Most of the alpha particles were undeflected.	The majority of the α-particles passed straight through the metal foil because they did not come close enough to any repulsive positive charge at all.
Some alpha particles were scattered by appreciable angles.	Only when a positive alpha particle approached sufficiently close to the nucleus, was it repelled strongly enough to rebound at high angles. Scattering was due to the mutual repulsion between the positive nucleus and the positive alpha particle. The small size of the nucleus explained the small number of alpha particles that were repelled in this way.
About 1 in 8000 alpha particles was 'back-scattered' through a very large angle indeed.	Back scattering occurs only when the incident alpha particle makes close to a "head-on" collision with a gold atom. Such collisions are quite rare because the nucleus of the atom is very small compared with the atom as a whole. Most of the atom is, in fact, empty space.

Rutherford concluded that all the positive charge and most of the mass of an atom formed an exceptionally small, dense core or nucleus. The negative charge consisted of a 'cloud of electrons' surrounding the positive nucleus.

Nuclear Radius

It was a challenging exercise for physicists to get some notion of the size of a nucleus by experiment. This is because the nucleus does not have a sharp edge; rather we should describe the edge of a nucleus as being 'fuzzy'. Nevertheless, physicists obtained an early idea of the size of a nucleus by firing α-particles at it. The distance of closest approach is an approximate upper limit of the size of a nucleus. Early experiments suggested that the nucleus might be considered to be a sphere of radius ≈ 10^{-15} m.

Variation of Nuclear Radius with Nucleon Number

Suppose we assume that the volume of a nucleon (ie, a proton or a neutron) in any nucleus is about the same. Then the **volume of the nucleus is directly proportional to the total number of nucleons** contained within it (that is, the **mass number** A). If we also suppose the nucleus to be spherical, then we can write that:

$r = r_0 A^{1/3}$

where r = the radius of a given nucleus (m)
r_0 = the constant of proportionality (m)
A = the mass number (number of nucleons within the nucleus)

The constant of proportionality, r_0, in the equation above is the radius of the nucleus for which $A = 1$, ie it is the radius of a proton. The value of r_0 is found by experiment to be **approximately** 1.2×10^{-15} m or 1.2 femtometres (sometimes written 1.2 fm or 1.2 fermi). However, it must be emphasised that the value of r_0 is not a well-defined constant and different values are obtained using different experimental techniques. The graph on the right shows the nuclear radius for various mass numbers.

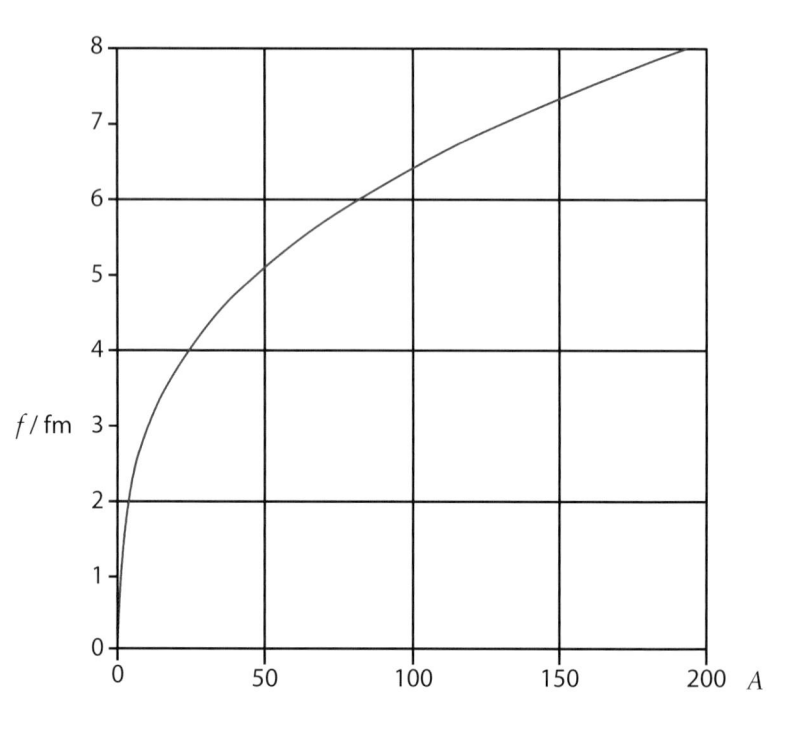

Nuclear Density

We can show that the density of nuclear matter is constant, ie that all atomic nuclei have the same density. Suppose a nucleus of mass number, A, is spherical with radius r and that the mean mass of its nucleons is m. Then its total mass M is given by:

$M = Am = \rho \dfrac{4\pi r^3}{3} = \rho \dfrac{4\pi r_0^3 A}{3}$ (since $r = r_0 A^{1/3}$)   where ρ = the density of nuclear matter (kg m^{-3})

Dividing both sides by A gives:
$m = \dfrac{4\pi r_0^3 \rho}{3}$

Hence the density of nuclear matter ρ is given by:
$\rho = \dfrac{3m}{4\pi r_0^3}$

Since m and r_0 are both constant values, the density of nuclear matter, r, must therefore also be constant. If we substitute the known values for m (the mass of the proton, 1.66×10^{-27} kg) and the value for r_0 (1.2×10^{-15} m) we can calculate the density of nuclear matter to be a staggering 2.3×10^{17} kg m^{-3}. The enormous density of nuclear matter in comparison with everyday matter is a reflection of the fact that in ordinary matter there is a great deal of empty space between the nucleus and the orbiting electrons. There is no empty space between the particles inside the nucleus.

4.5 THE NUCLEUS

Exercise 4.5

1. Your data and formulae sheet gives the equation for the radius of a nucleus as $r = r_0 A^{1/3}$.
   (a) In this equation what does the symbol A represent?
   (b) In terms of protons, neutrons and electrons, describe the structure of an atom of lead-208 ($^{208}_{82}Pb$).
   (c) Use the equation to find the radius of a lead-208 nucleus. Take $r_0 = 1.2$ fm.
   (d) Hence find the density of a lead-208 nucleus in kg m^{-3}. The mass of a neutron = 1.008665 u and the mass of a proton = 1.007276 u, where 'u' is a unit of mass equal to 1.66×10^{-27} kg.
   The volume of a sphere, $V = \dfrac{4\pi r^3}{3}$

2. In the α-particle scattering experiment, what feature of atomic structure explains the following observations?
   (a) The α-particles were deflected.
   (b) Most of the α-particles suffered a small deflection.
   (c) A small number of α-particles where almost deflected backward (ie by almost 180°).

3. Assuming the mean mass of a nucleon to be 1.66×10^{-27} kg, use the equations below to determine the density of the oxygen nucleus, $^{16}_{8}O$. Take $r_0 = 1.2$ fm.
   $r = r_0 A^{1/3}$
   The volume of a sphere, $V = \dfrac{4\pi r^3}{3}$

4.6 Nuclear Decay

Students should be able to:
4.6.1 Demonstrate an understanding of how the nature of alpha particles, beta particles and gamma radiation determines their penetration and range;
4.6.2 Calculate changes to nucleon number and proton number as a result of emissions;
4.6.3 Demonstrate an understanding of the random and exponential nature of radioactive decay;
4.6.4 Use the equation $A = -\lambda N$, where λ is defined as the fraction per second of the decaying atoms;
4.6.5 Use the equation $A = A_0 e^{-\lambda t}$, where A is the activity;
4.6.6 Define half-life;
4.6.7 Use the equation $t_{1/2} = \dfrac{0.693}{\lambda}$;
4.6.8 Describe an experiment to measure half-life of a radioactive source.

The Nucleus

The key features of atomic nuclei are as follows:

- Every atom has a central positively charged nucleus with a diameter of around 10^{-15} m.
- Atomic diameters are around 10^{-10} m, so the atom is typically 100 000 times bigger than its nucleus.
- Over 99.9% of the mass of an atom is in its nucleus.
- Atomic nuclei are totally unaffected by chemical reactions.
- Nuclei contain protons and neutrons. These are collectively known as **nucleons**.

The properties of nucleons are compared with those of the electron in the table below.

	Electron	Proton	Neutron
Relative Mass*	$1/1840$	1	1
Actual Mass**	9.109×10^{-31} kg	1.673×10^{-31} kg or $1836\, m_e$	1.675×10^{-31} kg or $1839\, m_e$
Relative Charge	-1	$+1$	0
Charge	-1.60×10^{-19} C	$+1.60 \times 10^{-19}$ C	Zero

* relative to the proton   ** m_e is the mass of the electron.

Nucleons are held together by the **strong nuclear force** which only acts over very short distances. The strong nuclear force is much stronger than the electric force of repulsion that exists between protons within the nucleus. Two key definitions are:

Atomic Number, Z   the number of protons in the nucleus.
Mass Number, A   the total number of nucleons in the nucleus.

A nucleus is described using these two numbers and the **chemical symbol of the element**. The general form is: $^{A}_{Z}X$
So, the uranium nucleus (chemical symbol U) containing 92 protons and 143 neutrons has $Z = 92$ and $A = 92 + 143 = 235$.
The symbol is therefore $^{235}_{92}U$.

Isotopes

Isotopes are nuclei with the **same** number of **protons** but **differing** numbers of **neutrons**. This means the value of the atomic number, Z, for isotopes of the same element must always be the same. So, for example, hydrogen has three stable isotopes, hydrogen (1p), deuterium (1p,1n) and tritium (1p, 2n). Their symbols are: $^{1}_{1}H$, $^{2}_{1}H$ and $^{3}_{1}H$.

Radioactivity

Some elements have unstable isotopes whose nuclei disintegrate (or decay) randomly and spontaneously. This effect known as **radioactivity**. The unit of activity is the Becquerel (Bq). **1 Bq is 1 disintegration per second**. Radioactive decay is both **random** (meaning that one cannot know precisely when the next decay will happen) and **spontaneous** (meaning that the decay of a nucleus is not caused by anything making it happen, ie it does not need something to trigger it). Radioactive sources can emit any of three different types of radiation, α-particles, β-particles or γ-rays.

Alpha (α) radiation

- An α-particle is a helium nucleus with two protons and two neutrons, and so has a mass number of 4.
- The symbol for an α-particle is $^{4}_{2}\alpha$ or $^{4}_{2}He$.
- α-particles are positively charged and so will be deflected in an electric or magnetic field.
- α-particles have poor powers of penetration and can only travel through about 4 cm of air.
- α-particles can easily be stopped by a sheet of paper.

4.6 NUCLEAR DECAY

- Since α-particles move relatively slowly and have a high momentum they **interact with matter producing intense ionisation** – a typical α-particle can produce about **100 000 ion–pairs per cm of air** through which it passes. Note: Ionisation is the process by which electrically neutral atoms or molecules are converted to electrically charged atoms or molecules (ions). It happens when an α-particle, β-particle or gamma ray causes an electron to be ejected from the atom or molecule. An ion-pair is the positively charged particle (positive ion) and the negatively charged particle (negative ion) simultaneously produced by an α-particle, β-particle or gamma ray interacting with the molecule.
- α-decay is described by the equation: $^{A}_{Z}X \rightarrow {}^{A-4}_{Z-2}Y + {}^{4}_{2}\alpha$

Beta (β) radiation

- A β-particle is a very fast electron and thus it has relative atomic mass of about $^{1}/_{1840}$.
- The symbol for a β-particle is $^{0}_{-1}\beta$ or $^{0}_{-1}e$
- β-particles are emitted from nuclei where the number of neutrons is much larger than the number of protons – one of the neutrons changes into a proton and an electron. The proton remains inside the nucleus but the electron is immediately emitted from the nucleus as a β-particle.
- The deflection of β-particles in an electric or magnetic field will be greater than that of α-particles, as the β-particles have a much smaller mass to charge ratio.
- β-particles interact less with matter than α-particles and have a greater penetrating power.
- β-particles can travel several metres in air, but can be stopped by 5 mm thick aluminium foil.
- β-particles have an ionising power between that of alpha and gamma radiation.
- The decay equation for β emission is: $^{A}_{Z}X \rightarrow {}^{A}_{Z+1}Y + {}^{0}_{-1}\beta$

Gamma (γ) radiation

- Unlike the other types of radiation, gamma radiation does not consist of particles.
- γ-rays are short wavelength, high energy electromagnetic waves emitted from unstable nuclei.
- The wavelength of γ-rays is characteristic of the nuclide that emits it, typically in the region 10^{-10} to 10^{-12} m.
- Like alpha and beta radiation, gamma radiation comes from a disintegrating unstable nucleus.
- γ-rays have no mass.
- Electric and magnetic fields have no effect on γ-rays.
- γ-rays have great penetrating power, travelling several metres in air.
- A thick block of lead or concrete is used to greatly reduce the effects of gamma radiation, but cannot stop it completely. A lead block about 5 cm thick will absorb around 90% of the γ-rays.
- Gamma radiation has the weakest ionising power as it interacts least with matter.
- The decay equation for gamma emission is: $^{A}_{Z}X^{*} \rightarrow {}^{A}_{Z}X + \gamma$

The asterisk is used to indicate that the parent nuclide is in an excited state.

Worked Example

(a) Complete the table below by inserting appropriate values of mass and charge for the alpha particle, the beta particle and gamma radiation.

	Relative mass (proton = 1)	Charge / C
Alpha particle		
Beta particle		
Gamma radiation		

(b) (i) How do these decay particles lose their kinetic energy after release into the atmosphere?
(ii) Explain why the alpha particle has a shorter range in air than the beta particle even though it is released with more kinetic energy.
(c) Part of the decay series of Uranium can be summarised as shown: $^{238}_{92}U \rightarrow (3\alpha + 2\beta) + {}^{A}_{Z}Ra$
Determine the values of A and Z for Radium (Ra).

(a)

	Relative mass (proton = 1)	Charge / C
Alpha particle	4	3.2×10^{-19}
Beta particle	$^{1}/_{1840}$	-1.6×10^{-19}
Gamma radiation	0	0

(b) (i) Collisions between the particles and the molecules of the medium through which they are passing cause energy to be transferred from the decay particle to the molecules of the medium.
(ii) The larger mass of the alpha particle means that it loses more momentum per collision.
(c) 3α results in mass number, A, decreasing by $3 \times 4 = 12$ and the atomic number, Z, decreasing by $3 \times 2 = 6$.
2β results in no change to the mass number and the atomic number, Z, increasing by 2.
So the overall change is that the mass number, A, decreases by 12 and the atomic number, Z, by 4 giving $^{226}_{88}Ra$.

Law of Radioactive Decay

The rate of decay of a particular nuclide is directly proportional to the number of unstable nuclei of that nuclide present. Therefore if there are N unstable nuclei present at time t then:

$A = -\lambda N$ where A = rate of disintegration with time (Bq)
$\phantom{A = -\lambda N\quad\text{where}\quad}$ N = number of unstable nuclei present
$\phantom{A = -\lambda N\quad\text{where}\quad}$ λ = a constant of proportionality called the decay constant (s^{-1}).

The minus sign is present because as t increases N decreases. The rate of disintegration of unstable nuclei is called **the activity** of the source and is measured in disintegrations per second or Becquerel (Bq).

Two mathematical consequences of this law are:

$A = A_0 e^{-\lambda t}$ and $N = N_0 e^{-\lambda t}$ where A_0 = the initial activity at time $t = 0$
$\phantom{A = A_0 e^{-\lambda t}\quad\text{and}\quad N = N_0 e^{-\lambda t}\quad\text{where}\quad}$ A = the activity at a time $t = t$
$\phantom{A = A_0 e^{-\lambda t}\quad\text{and}\quad N = N_0 e^{-\lambda t}\quad\text{where}\quad}$ N_0 = the original number of radioactive nuclei
$\phantom{A = A_0 e^{-\lambda t}\quad\text{and}\quad N = N_0 e^{-\lambda t}\quad\text{where}\quad}$ N = the number of radioactive nuclei at time $t = t$

Both the **activity, A, and the number of radioactive nuclei, N, decrease exponentially with time**.

Half–Life ($t_{½}$)

The half-life of a radioactive material is the time taken for half of the radioactive nuclei present to disintegrate. A mathematical consequence of this definition is that:

$$t_{½} = \frac{\ln 2}{\lambda} \approx \frac{0.693}{\lambda}$$

An equally acceptable definition for the half-life of a radioactive nuclide is **the time taken for the activity of that material to fall to half of its original value**.

> ### Worked Example
> *The half life of Americium-241 is 432 years. This isotope is commonly used in smoke alarms. Its activity in a particular smoke alarm is 35 kBq. Calculate the number of atoms of Americium-241 in this smoke alarm.*
>
> $t_{½} = \dfrac{0.693}{\lambda}$, so $\lambda = \dfrac{0.693}{t_{½}} = \dfrac{0.693}{432 \times 365 \times 86400} = 5.1 \times 10^{-11}\ s^{-1}$
>
> Activity $A = \lambda N$, so $N = A \div \lambda = 35 \times 10^3 \div 5.1 \times 10^{-11} = 6.86 \times 10^{14}$ nuclei and hence atoms of Americium-241.

Measuring the Half–life of a Radioactive Substance

The apparatus for this experiment is shown in the diagram below. The experiment involves the use of a Geiger-Müller (GM) tube and a counter to measure the activity of a sample of protactinium. When alpha, beta or gamma radiation enters the GM tube, it causes some of the argon gas inside to ionise and give an electrical discharge. This discharge is detected and counted by the counter. If the counter is connected to its internal speaker, you can hear the click when radiation enters the tube.

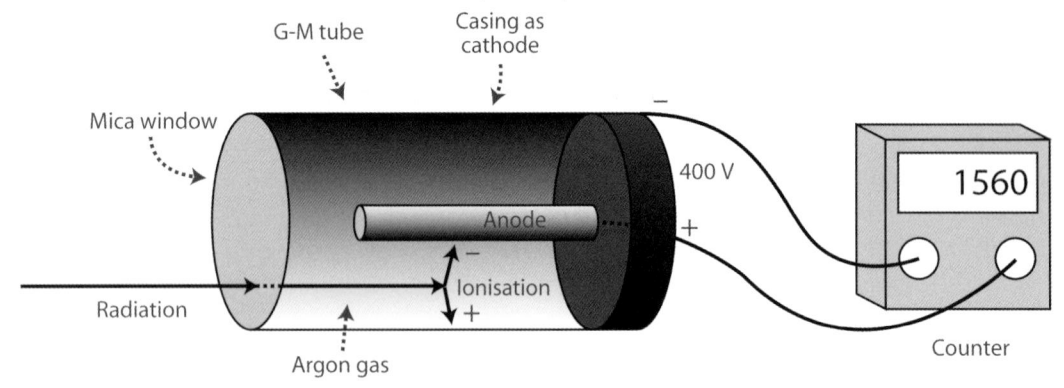

However, even in the absence of all known sources of radioactivity, the GM tube and counter still detects radiation. This is known as **background radiation** and it comes from the Sun, cosmic rays from space, hospital nuclear physics departments, nuclear power stations, granite rocks and so on. Before we use a GM tube to carry out any quantitative work on radiation we must first measure the background radiation if we are to correct for it in our experiment.

To Measure the Background Radiation
- First remove known sources of radiation from the laboratory, then set the GM counter to zero.
- Switch on the counter and start a stopwatch. After 30 minutes read the count on the counter.
- Divide the count by 30 to obtain the background count rate in counts per minute. A typical figure is around 15 counts per minute. This count must always be subtracted from any other count when measuring the activity from a specific source.

4.6 NUCLEAR DECAY

Setting up the Protactinium Source

Protactinium-234 is one of the decay products of uranium-238 and any compound of uranium-238 will have within it traces of protactinium. These traces may be conveniently extracted from it by chemical means. The protactinium decays by β-emission into another long–lived isotope of uranium (^{234}U) which is itself α–emitting. The very long half-life indicates low activity, which is not enough to interfere with this experiment. Moreover, the α–particles which are emitted will not penetrate a polythene bottle containing the protactinium. The β–activity at any instant of the extracted solution can therefore be used as a measure of the quantity of protactinium still present in it.

A thin-walled polythene bottle is filled with equal volumes of an acid solution of uranyl nitrate and pentyl ethanoate. When the liquids are shaken up together the organic ethanoate removes most of the protactinium present.

The solutions are not miscible and the protactinium remains in the upper layer when the liquids have once more separated.

The β-activity of the protactinium is observed with a GM tube and ratemeter, and the count rate is recorded at 10 second intervals.

Allowance is then made for the background count of the GM tube. If, for example, the measured rate with the GM tube and ratemeter is 32 counts per minute and the background rate is 15 counts per minute, then the corrected count rate is 32 – 15 = 17 counts per minute.

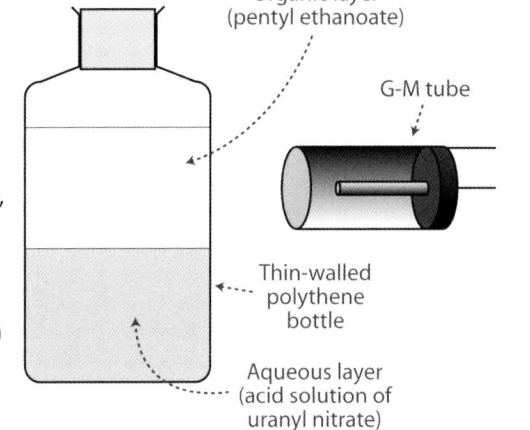

Treatment of the Results to Find the Half-Life of Protactinium

The corrected count rate of the protactinium is taken as a measure of its activity, A. By the Law of Radioactive Decay:

$A = A_0 e^{-\lambda t}$

Taking natural logs of both sides gives:

$\ln A = \ln A_0 - \lambda t$

Comparing this last equation with that for a straight line:

$y = c + mx$

we see that a graph of $\ln A$ (y-axis) against time (x-axis) is a straight line of gradient $-\lambda$ and y-axis intercept at $\ln A_0$.

We therefore plot a graph of $\ln A$ (y-axis) against time (x-axis), as shown on the right, draw the straight line of best fit and determine its gradient which is $(-\lambda)$.

We can then find the half–life by calculating the value of $\dfrac{0.693}{\lambda}$.

The generally accepted value for the half-life of protactinium-234 is 68 seconds.

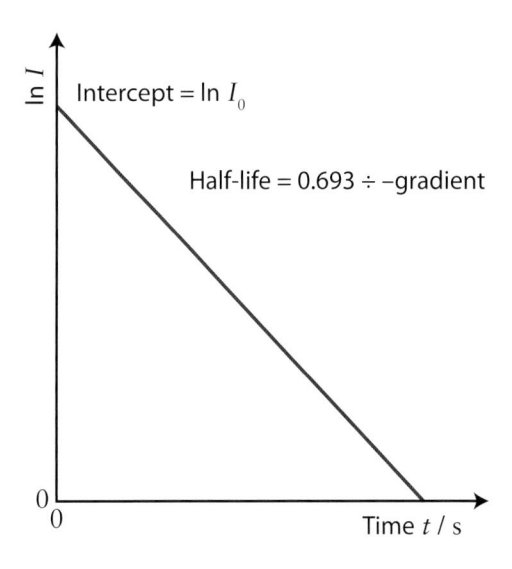

Exercise 4.6

1. The isotope of sodium $^{22}_{11}$Na is radioactive and emits gamma rays. It has a half life of 2.6 years. A mole of this isotope has a mass of 23.0 g.
   (a) Calculate the activity of 5.0 mg of this isotope.
   (b) How many unstable nuclei of this isotope remain after 2.0 years?

2. In an experiment to measure the half-life of a radioactive source measurements of the activity at various times was measured.
   (a) Describe how the corrected activity is obtained from the measurements.
   (b) Describe how a linear graph is obtained from the measurements and how the half-life is obtained from the linear graph.

3. ^{12}C and ^{14}C are isotopes of carbon. ^{12}C is stable but ^{14}C is radioactive and is used to estimate the age of wood found in an archaeological site. In any sample of wood the ratio of ^{14}C atoms to ^{12}C atoms is 1.0×10^{-12} and 12 g of ^{12}C contains 6.02×10^{23} atoms.
   (a) Calculate the number of ^{14}C atoms in 1 g of wood.
   (b) The half-life of ^{14}C is 5730 years. Calculate the initial activity of this 1 g of wood.
   (c) When the activity of the 1 g sample was measured it was found to be 0.09 Bq. Estimate the age of the sample of wood.

4.7 Nuclear Energy

Students should be able to:

4.7.1 Demonstrate an understanding of the equivalence of mass and energy;
4.7.2 Recall and use the equation $E = \Delta mc^2$ and demonstrate an understanding that it applies to all energy changes;
4.7.3 Describe how the binding energy per nucleon varies with mass number;
4.7.4 Describe the principles of fission and fusion with reference to the binding energy per nucleon curve.

Equivalence of Mass and Energy

In 1905 Albert Einstein published a remarkable paper which astonished the scientific community. It was called 'The Special Theory of Relativity'. For the purpose of this chapter, the important assertion made by Einstein was that **there is an equivalence to a mass, m, and energy, E,** given by the equation

$E = \Delta mc^2$ where E = energy (J)
m = change in mass (kg)
c = the speed of light in a vacuum (m s^{-1}) = 3×10^8 m s^{-1}

The equation $E = \Delta mc^2$ applies to **all** energy changes. If an object gains energy it is accompanied by an increase in mass, even if it is extremely small. The opposite applies in the case of a decrease of energy.

The Electron–Volt (eV) and the Unified Atomic Mass Unit (u)

The joule and the kilogram are much too large to be useful when dealing with atomic and nuclear processes. A much more appropriate unit for energy is the **electron-volt** (eV). The electron volt is the kinetic energy possessed by an electron accelerated from rest through a voltage of one volt.

For our purposes it is sufficient to know that:

1 eV = 1.6×10^{-19} J
1 MeV = 1 million electron-volts = 1.6×10^{-13} J

The unit commonly used for mass when dealing with atomic and nuclear processes is the **unified atomic mass unit** (u).

1 u = $\frac{1}{12}$ of the mass of the carbon-12 atom = 1.66×10^{-27} kg

Nuclear Binding Energy

The mass of a nucleus is **always less** than the sum of the masses of its constituent nucleons. This difference in mass is called the **mass defect** of the nucleus.

Mass defect = total mass of the nucleons – mass of the nucleus

This reduction in mass arises due to the act of combining the nucleons to form the nucleus. When the nucleons are combined to form a nucleus a tiny portion of their mass is converted to energy. This energy is called the **binding energy** of the nucleus. Binding energies are usually quoted in MeV. The binding energy of a nucleus is the amount of energy that has to be **supplied** to separate the nucleons completely, ie to an infinite distance apart. Binding energies can be given in joules (J), but are usually quoted in millions of electron volts (MeV).

> **Worked Example**
>
> *The electrons in an X-ray tube are accelerated through a potential difference of 100 kV.*
> *Calculate the increase in mass of an electron.*
>
> 1 electron-volt is the kinetic energy possessed by an electron accelerated through 1 volt. So for 100 kV:
>
> Energy = $E \times V$ = $1.6 \times 10^{-19} \times 1.0 \times 10^5$ = 1.6×10^{-14} J
>
> We now need to find the increase in mass resulting from this additional energy:
>
> $E = \Delta mc^2$, so $\Delta m = \dfrac{E}{c^2} = \dfrac{1.6 \times 10^{-14}}{(3 \times 10^8)^2} = 1.78 \times 10^{-31}$ kg

The **binding energy per nucleon** is a useful measure of the stability of a nucleus. The average binding energy per nucleon varies with nucleon number as shown at the top of the next page. A graph of average binding energy per nucleon against atomic number has a similar shape.

4.7 NUCLEAR ENERGY

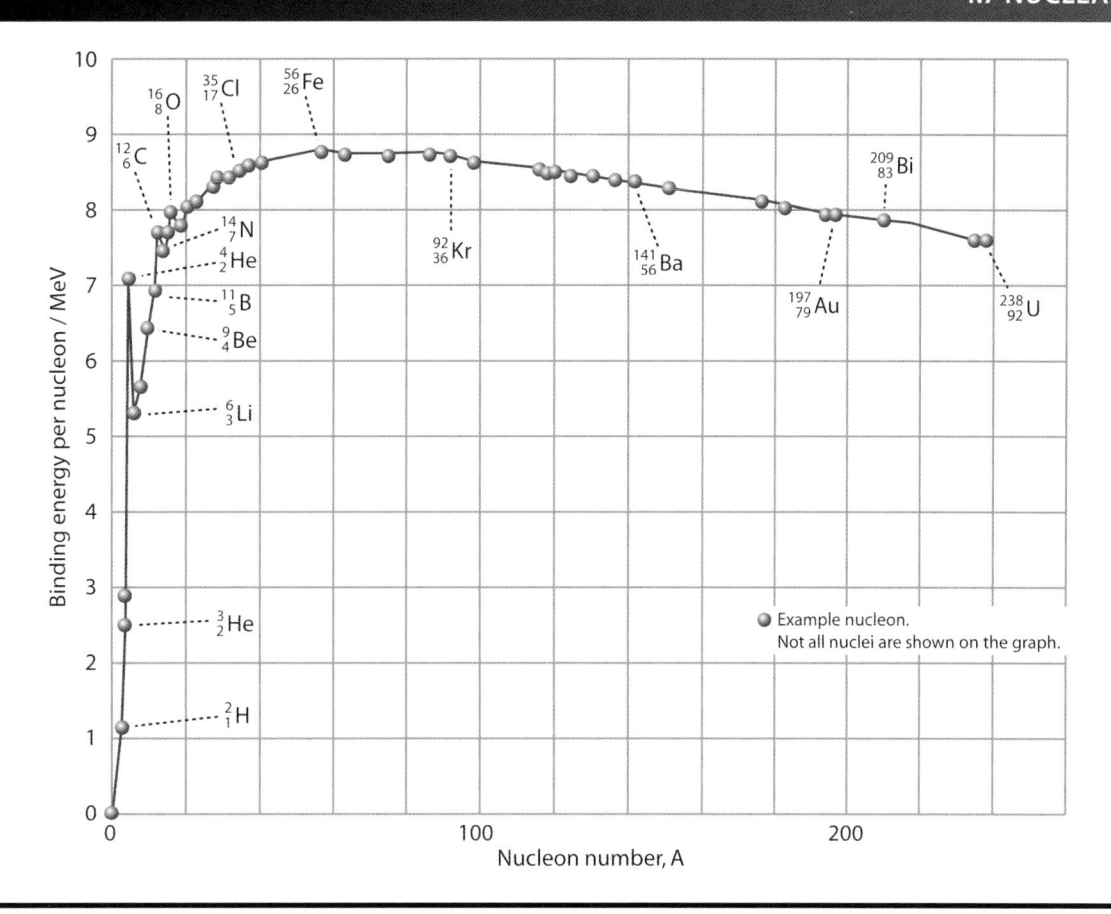

Worked Example

Calculate (a) the binding energy and (b) the binding energy per nucleon of helium-4. Give your answers in MeV. The masses of the neutron and proton are 1.0087 u and 1.0078 u respectively. The mass of the helium nucleus is 4.0026 u.

(a) The He-4 nucleus contains 2 protons and 2 neutrons.
   mass of 2 protons $= 2 \times 1.0078$ u $= 2.0156$ u
   mass of 2 neutrons $= 2 \times 1.0087$ u $= 2.0174$ u
   mass of the constituent nucleons $= 4.0330$ u
   mass of the helium nucleus $= 4.0026$ u
   Therefore, mass defect $= 4.0330 - 4.0026 = 0.0304$ u. Convert this to kg $= 0.0304 \times 1.66 \times 10^{-27} = 5.046 \times 10^{-29}$ kg
   Binding energy $= \Delta mc^2 = 5.046 \times 10^{-29} \times (3 \times 10^8)^2 = 4.54 \times 10^{-12}$ J
   However, we have to convert the unit from joules to MeV, so:
   Binding energy $= 4.54 \times 10^{-12} \div 1.6 \times 10^{-13} = 28.39$ MeV.

(b) There are 4 nucleons in the He-4 nucleus. So:
   Binding energy per nucleon = binding energy ÷ no. of nucleons $= 28.38 \div 4 = 7.1$ MeV per nucleon.

Nuclear Fission

Fission is the division of a massive nucleus into two less massive nuclei, each with a higher binding energy (BE) per nucleon. Nuclei to the right of the peak of the 'BE per nucleon' against 'nucleon number' curve (as shown on the right) undergo fission to increase stability by reaching a higher BE per nucleon.

The total binding energy of these fission fragments is higher than that of the heavy nucleus. Due to the increase in the total binding energy, some of the mass of the heavy nucleus is converted to kinetic energy of the fission fragments. For example, uranium-235 undergoes fission when it absorbs a neutron, as shown in the lower diagram on the right.

Neutrons produced in the fission process can go on to cause further fissions. This is called a **chain reaction**.

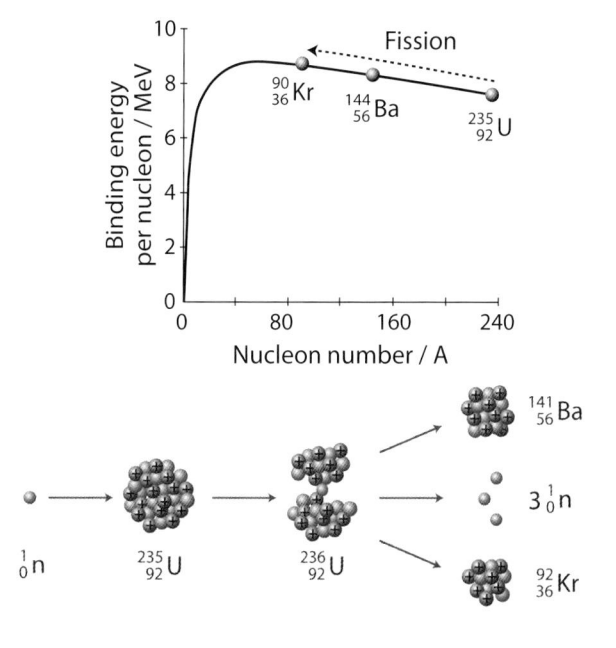

PHYSICS FOR CCEA A2 REVISION GUIDE, 2ND EDITION

Nuclear Fusion

Fusion is the joining of lighter nuclei to produce a heavier and more stable nucleus. Nuclei to the left of the peak of the 'BE per nucleon' against 'nucleon number' curve (as shown on the right) undergo fusion to increase stability by reaching a higher BE per nucleon.

The fusion process results in the release of energy since the average binding energy of these fusion products is higher than that of the lighter nuclei which join together.

In fusion some of the mass of the lighter nuclei is converted to kinetic energy of the fusion product. This means that the mass of the heavier nucleus is less than the total masses of the two light nuclei that fuse.

The reaction cannot take place at room temperature because of the repulsive electric force between the positively charged nuclei. It only occurs when the speed of the colliding nuclei is great enough for the nuclei to overcome this repulsive force and they can come close enough for the attractive, but very short range, strong nuclear force to cause fusion to occur.

For example, hydrogen nuclei can fuse, as shown in the lower diagram on the right, only when the temperature is high enough (about 15 million degrees Celsius).

At such high temperatures matter exists in a fourth state known as **plasma**. The atomic electrons break free from the nucleus, and the gas–like fluid is a mixture of electrons, positive ions and free nuclei.

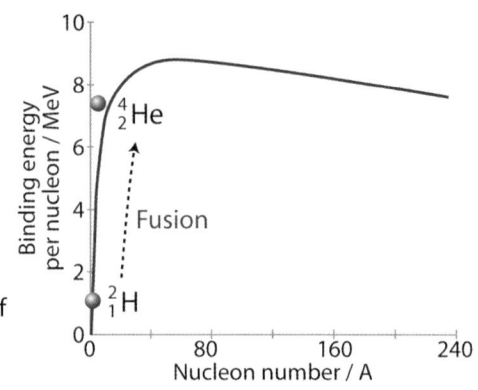

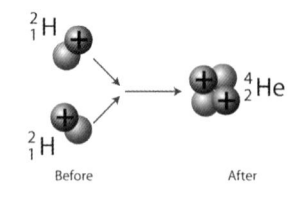

Exercise 4.7

1. (a) Calculate the increase in mass of an electron when it is accelerated from rest through a potential difference of 1 MeV. Remember energy change = charge × potential difference.
   (b) A car battery is charged at 13.5 V, 8 A for 5 hours. Calculate the increase in mass of the car battery.

2. Calculate the total binding energy and mean binding energy per nucleon of iron, $^{56}_{28}$Fe .
   Mass of iron = 55.934938 u
   Mass of a neutron = 1.008665 u
   Mass of a proton = 1.007276 u

3. One example of the fission of U-235 is the following reaction:
   $^{235}_{92}$U + $^{1}_{0}$n → $^{140}_{55}$Cs + $^{93}_{37}$Rb + 3$^{1}_{0}$n + energy

   Calculate the amount of energy released, in joules, in this reaction.
   Mass of U-235 atom = 235.04394 u
   Mass of Rb-93 atom = 92.92204 u
   Mass of Cs-140 atom = 139.91728 u
   Mass of a neutron = 1.008665 u

4. Calculate the energy released in the following fusion reaction:
   $^{3}_{1}$H + $^{2}_{1}$H → $^{4}_{2}$He + $^{1}_{0}$n
   Mass of $^{3}_{1}$H = 3.01605 u
   Mass of $^{2}_{1}$H = 2.01410 u
   Mass of $^{1}_{0}$n = 1.00866 u
   Mass of $^{4}_{2}$He = 4.00260 u

4.8 Nuclear Fission and Fusion

Students should be able to:

4.8.1 Demonstrate an understanding of the terms chain reaction, critical size, moderators, control rods, cooling system and reactor shielding, as used in describing a fission reactor;

4.8.2 Demonstrate an understanding of the social, environmental, security and economic issues surrounding the use of nuclear power as a solution to a future energy crisis;

4.8.3 Describe the ITER (tokamac concept) fusion reactor in terms of fuel, D-T reaction, temperature required, plasma, three methods of plasma heating, vacuum vessel, blanket, magnetic confinement of plasma, difficulties of achieving fusion on a practical terrestrial scale, and advantages and disadvantages of fusion;

4.8.4 Describe the following methods of plasma confinement: gravitational, inertial and magnetic.

Nuclear Fission

The common features of the **fission** processes are:
- It always releases huge quantities of energy, about 80% of which is carried away by the kinetic energy of the two major fission fragments. Burning one atom of carbon (eg, in the form of coal) typically releases 5 eV. By contrast, the fission of one uranium nucleus releases more than 200 million eV.
- The fission fragments are often radioactive and their subsequent decay accounts for approximately a further 10% of the total energy released.
- Extremely penetrating and highly dangerous gamma rays are always produced along with the fission fragments. These gamma rays and the kinetic energy of the sub-atomic particles produced account for the remaining 10% energy released.

In nuclear fission, further neutrons are always produced, on average around 2.5 per fission. These additional neutrons can go on to produce further fission events, yielding more neutrons which produce even more fission events and so on. In this way fission has the potential to produce a chain reaction. An **uncontrolled** chain reaction is what takes place in an atomic bomb, releasing enormous amounts of energy in a very short time. However, the focus of this chapter is how to bring about a **controlled** chain reaction in a fission **reactor**, for example to generate electricity.

Fission Nuclear Reactors

There are **four common types** of fission nuclear reactor: the Magnox type, the Advanced Gas-cooled Reactor (AGR), the Pressurised Water Reactor (PWR) and the Fast Reactor. The UK's oldest nuclear reactors were of the Magnox type, so-called because the natural uranium fuel was clad in a tube made of magnesium alloy. AGRs use circulating gas, almost invariably carbon dioxide, as the coolant. The 'coolant' is the material used to transfer energy away from the nuclear reactor to generate electricity. PWRs use water as the coolant, under such high pressure that even at a temperature of over 200°C it remains liquid. Fast reactors use plutonium rather than uranium as the fuel and liquid sodium as the coolant. Fast reactors tend to be used in nuclear-powered submarines. Because the AGR is now the most common type of reactor used in the UK, the remainder of our discussion shall be limited to that reactor type.

Nuclear Fuel and Moderators

Natural uranium consists of about 99.3% uranium-238 and 0.7% uranium-235. Uranium-238 is fissile (meaning, it can undergo nuclear fission), but only with very fast neutrons. Conversely, uranium-235 is fissile only with slow neutrons (known as thermal neutrons). The neutrons emitted in the fission of uranium-235 are too slow to cause fission in uranium-238, but need to be slowed down even further to cause further fission in uranium-235. This is achieved by the use of a material called a moderator, of which the most common types are graphite, water (H_2O) or heavy water (deuterium oxide or D_2O).

To make the uranium within the reactor more likely to undergo fission with slow neutrons it needs to be **enriched** by raising the proportion of fissile uranium-235 from about 0.7% in the natural ore to about 3% in the nuclear fuel rods.

Design of a Nuclear Reactor

The AGR is an example of a **graphite moderated reactor**. The enriched uranium is in long, sealed tubes, called **fuel rods** which are arranged inside a block of graphite. The neutrons released by the fission of uranium-235 collide inelastically with the atoms of the graphite moderator and are slowed down to a speed where they are more likely to cause further fission in uranium-235 than be absorbed by the uranium-238. A schematic diagram of an AGR nuclear fission reactor is shown overleaf.

Critical Size

When a fission event occurs, there are three possible fates for the fission neutrons produced:
- they might escape from the fuel rod without causing a further fission;
- they might be absorbed by a neighbouring nucleus, again without causing fission;
- they cause another fission event in a uranium nucleus.

For the reaction to be sustained, on average at least one of the neutrons released by each fission event must go on to produce a further fission. The bigger the size of the uranium fuel assembly, the more likely it is that a fission neutron will go on to produce another fission event.

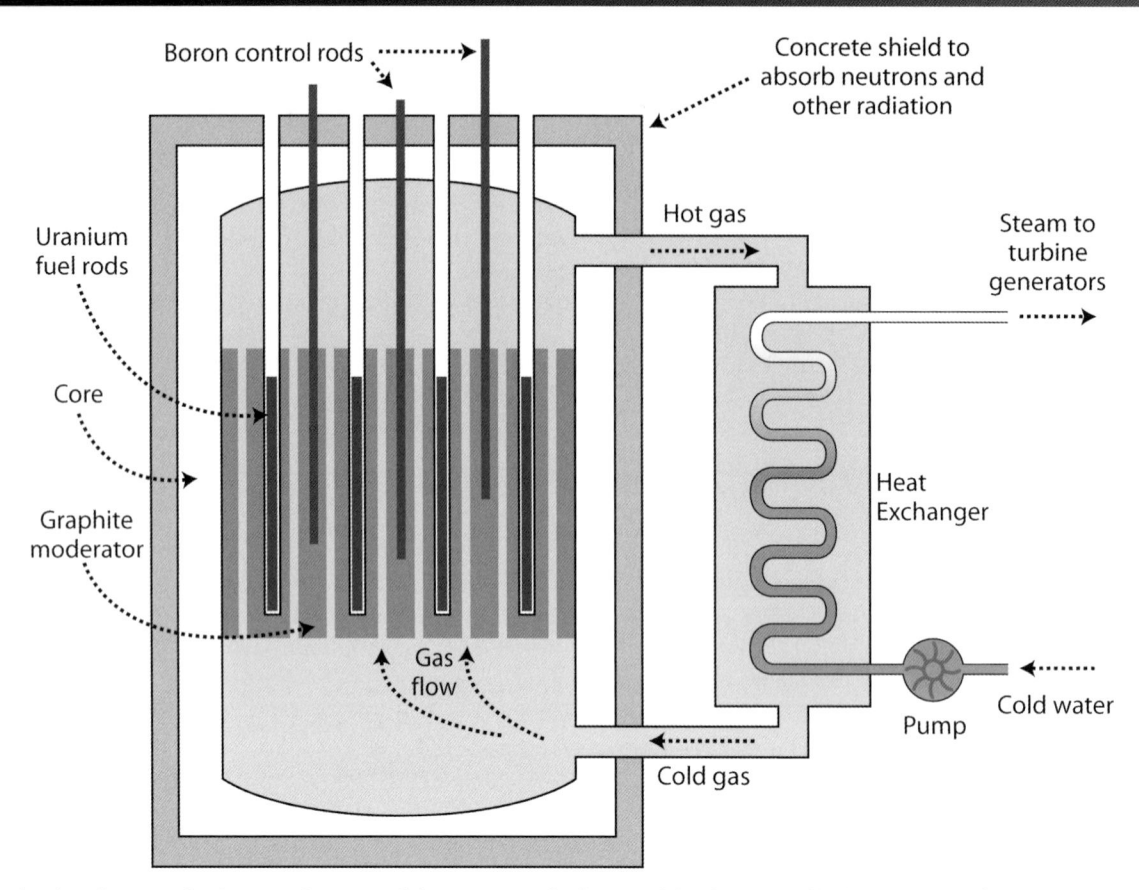

We can define the **critical size** of the fuel assembly as that which is just capable of sustaining a chain reaction within it. Below the critical size, too many of the neutrons which might have induced further fission escape and the chain reaction dies away. A typical nuclear power station has a fuel assembly which is around 5% above the critical size.

Control Rods
If, on average, much more than one of the neutrons produced by fission went on to cause further fission, then the reaction would quickly go out of control. So, **boron–coated steel rods, called control rods, are used to absorb excessive neutrons** before they can cause fission. When the control rods are lowered into the reactor the number of available neutrons is decreased and fewer fission events can occur.

Coolant
The heat energy produced by the fission reaction is removed by passing a coolant through the reactor. This coolant then passes its energy to water by flowing through a **heat exchanger**, which produces steam that then drives the **turbines** which turn the **electricity-producing generators**. The part of a nuclear power station associated with the turbines and generators is exactly the same as would be seen in a conventional power station burning fossil fuels – the only difference is the source of the energy used to produce the steam.

Reactor Shielding
Every civil reactor is surrounded by a very thick concrete shield to prevent potentially dangerous radiation, particularly very penetrating gamma waves and neutrons, from reaching workers and the wider community.

Nuclear Fusion

Fusion in the Stars
Almost all of the energy we receive on Earth comes from the Sun as a result of nuclear **fusion**. All the elements which form the basis of our material world as well as the material in all living things on the Earth were formed by fusion in stars. The Sun consists mainly of hydrogen and helium. At the core of the Sun the temperature is many millions of kelvins resulting in a constant fusion of hydrogen nuclei. The reactions can be summarised by:

$$^1_1H + {}^1_1H + {}^1_1H + {}^1_1H \rightarrow {}^4_2He + \text{other products} + \text{energy}$$

To overcome the electrostatic repulsion between nuclei, they need to have sufficient kinetic energy. This requires temperatures of the order of 10^8 or 10^9 K. At these temperatures all matter exists as plasma.

4.8 NUCLEAR FISSION AND FUSION

Estimate of the Temperature Required for Fusion

In nuclear fusion, two light particles – say two protons – are brought sufficiently close together so that they fuse to form a more massive particle. However, protons are positively charged which means that as two protons come closer they repel each other. The closer they get the greater the repulsive force. For fusion to take place, they must get very close, certainly closer than the diameter of the nucleus (about 1×10^{-15} m). If the protons are projected towards each other the repulsion between their positive charges causes their kinetic energy to decrease and their potential energy to increase. The energy needed to bring a pair of protons sufficiently close to bring about fusion is about **110 keV**. We can use this information to calculate the temperature to which the proton assembly must be heated.

Recall from chapter 4.2 that the temperature of a gas is a measure of the mean kinetic energy of the gas molecules and that:

Mean kinetic energy = $\frac{3}{2}m<c^2> = \frac{3}{2}kT$ where m = the mass of each particle (kg)
$<c^2>$ = the mean square speed (m² s⁻²)
k = the Boltzmann constant (1.38×10^{-23} J K⁻¹)
T = temperature (K)

If we treat the collection of protons as a gas, of mean kinetic energy 110 keV (where 1 keV = 1.6×10^{-16} J), then:

$$T = \frac{2 \times \text{mean kinetic energy}}{3 \times k} = \frac{2 \times 110 \times 1.6\times10^{-16}}{3 \times 1.38\times10^{-23}} = 850\times10^6 \text{ K}$$

Plasma Confinement

As we have seen, fusion on a practical terrestrial scale requires a temperature of 850 million kelvins. There is no possibility of using a conventional reaction vessel as its walls would simply vaporise. There are three possible methods to confine the plasma long enough for fusion to occur. These are summarised in the below table:

Gravitational Confinement	Inertial Confinement	Magnetic Confinement
• Gravitational forces in stars can provide the plasma confinement. • The inward gravitational pull balances the outward forces created by the pressure of the plasma and radiation pressure (photons impacting on the particles of the gas plasma and exerting an outward pressure). • Gravitational confinement would not work on the Earth since we require an enormous mass of material to provide gravitational forces strong enough to balance the forces tending to dissipate the plasma.	• Inertial confinement involves using intense ion or laser beams directed at a solid fuel pellet (such as lithium hydride). • The laser beams provide the energy to heat the material to the temperature required for fusion. • The idea is to produce fusion for long enough to extract the energy before the plasma escapes: rather like pulling a table-cloth away from a table before the tea-cups fall over! • The fusion seen in this situation is of the type: $^{6}_{3}\text{Li} + ^{2}_{1}\text{H} \rightarrow 2\,^{4}_{2}\text{He}$	• Magnetic confinement uses magnetic fields to hold the plasma. • The magnetic field is produced by electric current flowing in a coil wound into a shape known as toroid (doughnut). • The magnetic field produced is circular within the highly evacuated toroidal chamber. • The charged particles in the plasma are moving in a magnetic field and experience a force. • The force causes charged plasma particles to circulate endlessly in helical paths around the magnetic field lines produced by water–cooled toroidal field coils.

The Deuterium-Tritium (or D-T) Reaction

The fusion process with greatest promise at the moment is known as the **deuterium-tritium (or D-T) reaction:**

$^{2}_{1}\text{H} + ^{3}_{1}\text{H} \rightarrow ^{4}_{2}\text{He} + ^{1}_{0}\text{n} + 17.6 \text{ MeV}$

Suppose that surrounding the reactor is a blanket of lithium. Lithium can absorb the fusion neutron and then fission according to the reaction:

$^{7}_{3}\text{Li} + ^{1}_{0}\text{n} \rightarrow ^{4}_{2}\text{He} + ^{3}_{1}\text{H} + ^{1}_{0}\text{n}$

The neutrons released in both reactions above can go on to sustain a chain reaction in lithium, thus converting the lithium into tritium fuel to continue the D-T process. Once this reaction can be sustained, the heat from the kinetic energy given to the helium nuclei would be used to heat water to produce steam, which then would be used to drive turbines and hence generate electricity.

Advantages of Nuclear Fusion

Seeking to achieve nuclear fusion by the D-T process is attractive because:
- Supplies of fuel are readily available and virtually inexhaustible. Sea water contains 1 atom of deuterium for every 5000 of ordinary hydrogen and this can be extracted by electrolysis. Tritium can be created from lithium.
- There is none of the toxic and highly radioactive waste associated with fission. There are waste products in the form of materials that have been irradiated with neutrons, but these pose much less of a problem than the fission fragments

associated with fission of uranium which have a very long half-life.
- There is a greater yield of energy per kilogram of fuel consumed from a hydrogen fusion reactor than from a fission reactor. It has been estimated that a 2000 MW fusion power station would require only 0.5 kg of deuterium and 1.8 kg of lithium per day.
- Fusion reactors are inherently 'fail safe' because the fuel is continuously fed into them. If this feed is stopped, then the reaction ceases. In a fission reactor all the fuel is place in the reactor to start with as there has to be a least a critical mass before the fission chain reaction can begin.

Disadvantages of Nuclear Fusion
- It requires a large initial energy input before useful output is obtained.
- The process is still unproven as a means of providing electricity on a commercial scale. To date the process is not close to achieving the aim of producing more energy than is consumed, and it has been estimated that this will not happen until at least 2050.
- Commercial power plants would be extremely expensive to build. The highest cost is likely to be the need for large superconducting magnets to providing the magnetic containment of the plasma.

ITER – International Thermonuclear Experimental Reactor

The CCEA specification requires students to be able to describe elements of **the International Thermonuclear Experimental Reactor** (**ITER**) a fusion reactor that is currently being developed in France in a collaboration of 35 countries. The picture on the right shows a cross-section of how ITER will look when completed.

At the centre of the complex is a tokamak vacuum vessel, used to contain the plasma in which the fusion reaction takes place. The walls of the vacuum vessel are lined with beryllium blankets approximately 0.5 m thick with a mass of almost 5000 kg. These blankets will provide shielding from the high-energy neutrons produced by the fusion reactions.

As the neutrons are slowed in the blanket, their kinetic energy is transformed into heat energy and collected by the water coolant. In a fusion power plant, this energy would be used for electrical power production. Cooling water circulating through the vessel's double steel walls will remove the heat generated during operation, producing steam to drive turbines to produce electricity.

The reactor uses the deuterium-tritium reaction discussed in the previous section, ie using two isotopes of hydrogen, deuterium($^2_1$H) and tritium ($^3_1$H).

ITER uses three methods to heat the plasma to the very high temperature (150 million °C) needed for fusion:
- The plasma is a good conductor of electricity. The changing magnetic fields that are used to contain the plasma also produce a very large current by electromagnetic induction. This current passes through the plasma causing the electrons and ions to gain kinetic energy and collide. These collisions result in a resistance (similar to what happens when a metal conducts electricity) and produce a heating effect. However, as is the case with metals, the resistance of a plasma increases with its temperature. As the resistance increases, the current decreases and so the heating effect decreases, which means that further sources of heating are then required.
- A beam of deuterium ions is accelerated by an electric field to a high velocity. However, before the ions enter the plasma they pass through another electric field and as a result gain an electron to make them electrically neutral. These high velocity neutral particles collide with the ions and electrons in the plasma and transfer their energy to the plasma particles, causing further heating.
- The final step is to direct high frequency microwaves into the plasma. These waves transfer their energy to the particles in the plasma. To ensure the energy is transferred as efficiently as possible, three specific frequencies of microwave are used, each one being matched to a specific type of ion or electron within the plasma.

It is expected that the ITER fusion reactor will produce around 500 MW of power, about the same output as a coal-fired power plant. However, ITER is first and foremost a large scale physics project, whose aim is to develop a reliable fusion reactor, and is not designed to produce electricity commercially.

4.8 NUCLEAR FISSION AND FUSION

Nuclear Energy, Society and the Environment

It is known that carbon dioxide in the Earth's atmosphere traps solar energy, by radiating infra-red radiation emitted from the planet's surface back to the Earth. Increasing levels of atmospheric carbon dioxide is causing global temperatures to rise, leading to harmful climate change. A significant source of carbon dioxide is the burning of fossil fuels to generate electricity. An increased use of nuclear energy has been proposed as a solution to the problem of carbon dioxide emissions.

Nuclear power stations do not release carbon dioxide into the atmosphere. It is true that when the whole process is taken into account, including the mining, refining and recycling of uranium fuel rods, the amount of carbon dioxide released is many times more than that produced from renewable energy sources. However, it is still many times less than burning fossil fuels.

Fossil fuels are also finite resources and will eventually run out, with oil likely to run out first. Due to our reliance on fossil fuels, many therefore foresee a future **energy crisis**. In order to generate sufficient power to replace fossil fuels, nuclear supporters argue that we must expand all other forms of energy production, including both renewable energy sources and nuclear power. Opponents of nuclear energy argue that the environmental, health and security risks make it unsuitable as a replacement for fossil fuels. They argue that funds should be diverted away from nuclear power to develop renewable energy sources do not pose such risks. Some issues with nuclear power are:

- Nuclear energy (from fission) produces large quantities of toxic, radioactive waste and that must be stored safely and securely for between 10 000 years and 240 000 years in order to prevent health and environmental disasters from radioactive contamination. Finding safe storage facilities for this waste poses problems for many countries. Many would not welcome such storage facilities in their neighbourhood. One argument for ongoing research into nuclear fusion is that it does not produce waste products like this.
- Nuclear power creates employment opportunities for many people in the local area. However, those living close to nuclear power plants and radioactive waste storage sites have many concerns.
- Nuclear power is a secure source of energy provided the countries that provide the uranium ore have stable governments and societies. However, some sources of uranium ore are in countries that might be considered to be less stable. These sources of uranium will become more important in the future as supplies from stable sources become depleted.
- Some of the technology used for nuclear power can also be used to produce nuclear weapons.

Exercise 4.8

1. A simplified diagram of a nuclear fission reactor is shown below. Name the parts labelled 1 to 4 and describe the role they play in the reactor.

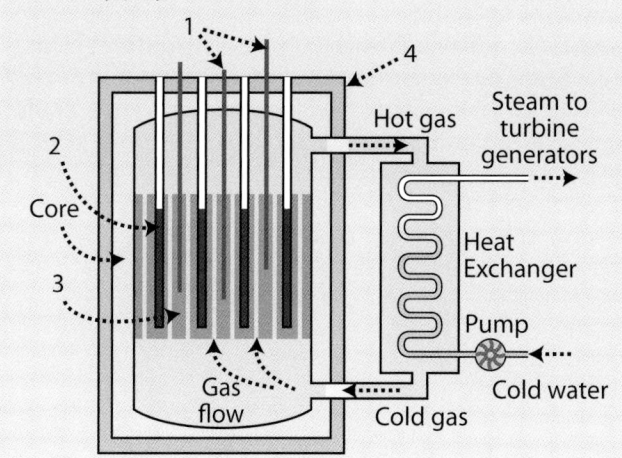

2. When uranium nuclei undergo fission, neutrons are released. State three things that can happen to these neutrons.

3. 'Critical size' is a term used in connection with nuclear reactors. Explain the meaning of this term.

4. Nuclear fusion could replace fossil fuels as an energy resource on the Earth. One reaction which could lead to the release of energy is the fusion of deuterium and tritium (the D-T reaction).
   (a) State the equation for the D-T reaction.
   (b) Give two reasons why this reaction is most suitable for terrestrial fusion.

Unit 5 (A2 2):
Fields and their Applications

5.1 Force Fields

Students should be able to:

5.1.1 Explain the concept of a field of force, using field lines to describe the field, indicate its direction and show the field strength.

Fields of Force

A **field of force** is a region of space within which objects with a particular property experience a force.
- A **gravitational field** is a region of space within which a mass will experience a force.
- An **electrical field** is a region of space within which a charge will experience a force.
- A **magnetic field** is a region of space within which a moving charge will experience a force.

Around every charge there exists an electric field, around every mass there exists a gravitational field and around moving charges (or an electric current) there exists a magnetic field.

Field Lines

Field lines give us a visual picture of a field. The direction of the field at any point is given by the tangent to the field line, in the direction of the arrow, at that point. At any given point, the field has only one direction. It therefore follows that field lines cannot cross, because if they did it would mean that the field had more than one direction at that point, which is a contradiction.

There should always be an **arrow** on a field line to show the direction of the field. The closer the field lines are drawn together, the greater is the strength of the field. Field lines which are parallel and equally spaced are called **uniform** fields. This means the strength of the field is constant everywhere within the field. The diagrams below show typical field lines for three common types of field: the radial (non-uniform) gravitational field around the Earth; the radial (non-uniform) electric field around isolated electrical charges; and the non-uniform magnetic field around a bar magnet.

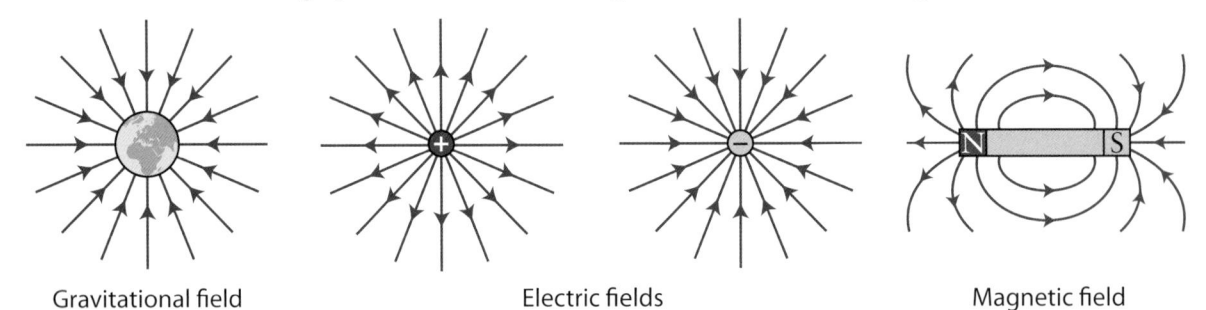

Gravitational field   Electric fields   Magnetic field

5.2 Gravitational Fields

Students should be able to:

5.2.1 define gravitational field strength;

5.2.2 recall and use the equation $g = \dfrac{F}{m}$;

5.2.3 state Newton's law of universal gravitation;

5.2.4 recall and use the equation for the gravitational force between point masses, $F = G\dfrac{m_1 m_2}{r^2}$;

5.2.5 recall and apply the equation for gravitational field strength, $g = \dfrac{Gm}{r^2}$, and use this equation to calculate the mass, m;

5.2.6 apply knowledge of circular motion to planetary and satellite motion;

5.2.7 show that the mathematical form of Kepler's third law (T^2 proportional to r^3) is consistent with Newton's law of universal gravitation;

5.2.8 demonstrate an understanding of the unique conditions of period, position and direction of rotation required of a geostationary satellite.

Definition

In a gravitational field, the **gravitational field strength**, g, at a point is equal to the force which would be produced on a test mass of 1 kg at that point, ie:

$g = \dfrac{F}{m}$ where g = gravitational field strength (N kg^{-1})
F = force (N)
m = mass (kg)

Field strength is a vector, ie it has both magnitude and direction.

Newton's Law of Universal Gravitation

Newton's Law of Universal Gravitation states that **between every two point masses there exists an attractive gravitational force which is directly proportional to the mass of each and inversely proportional to the square of their separation**. This is written mathematically as:

$F = G\dfrac{m_1 m_2}{r^2}$ where: m_1 and m_2 = the respective point masses (kg)
r = the distance between them (m)
G = a constant (the universal gravitational constant) which is equal to 6.67×10^{-11} N m^2 kg^{-2}

Gravitational Field Strength

Field lines

The direction of a gravitational field line at a point shows the direction of the gravitational force on a mass of 1 kg at that point. For a uniform sphere the gravitational field pattern is described as **radially inwards**, because all field lines appear to converge at the centre of mass of the sphere (as shown on the near right).

For a person on the surface of the Earth the field lines strike the surface at right angles (far right). The radius of the Earth is so large that the field lines appear parallel and equally spaced. Such a field is called a **uniform field**.

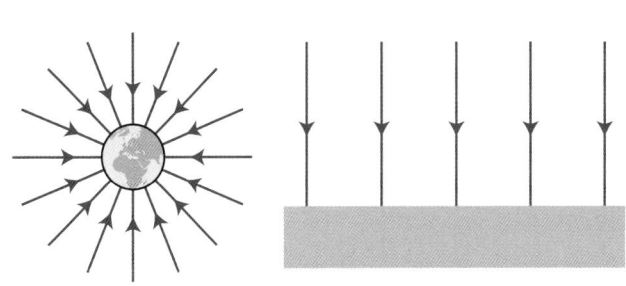

Variation of gravitational field strength with height above the Earth's surface

Gravitational field strength at a point is equal to the force which would be produced on a test mass of 1 kg at that point. The diagram on the right shows a point mass M. The gravitational force on a mass of 1 kg at the point X is given by:

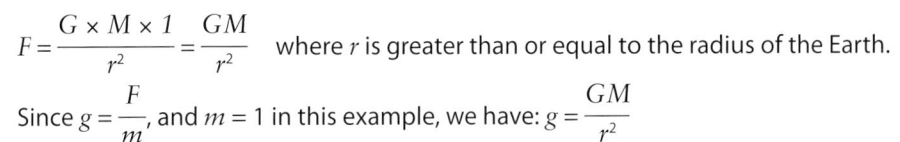

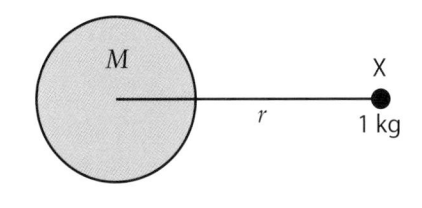

$F = \dfrac{G \times M \times 1}{r^2} = \dfrac{GM}{r^2}$ where r is greater than or equal to the radius of the Earth.

Since $g = \dfrac{F}{m}$, and $m = 1$ in this example, we have: $g = \dfrac{GM}{r^2}$

Gravitational field strength is a vector and has the units N kg^{-1} which is equivalent to m s^{-2}.

41

The **weight** of this 1 kg object is due to the gravitational attraction of the Earth. When the object is at the Earth's surface:

$$W = mg = \frac{G \times M_E \times 1}{R_E^2}$$ where M_E = the mass of the Earth (kg)
R_E = the radius of the Earth (m)

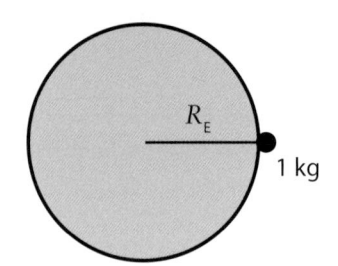

Calculating the mass of the Earth

The mean value of g over the surface of the Earth is generally taken as 9.81 m s^{-2}. The radius of the Earth is around 6400 km. Substitution of values for G and R_E into the above equation gives a mass M_E of 6×10^{24} kg.

Planetary Motion and Kepler's Third Law

Johannes Kepler was a seventeenth century German mathematician and astronomer. He defined three laws relating to the motion of the planets around the Sun. One of them, the third law, states that **the square of the period of revolution of the planets about the Sun is directly proportional to the cube of their mean distances from it**.

The CCEA specification requires students to be able to demonstrate that Kepler's third law is consistent with Newton's law of universal gravitation. The derivation below must therefore be memorised.

The orbit of a planet around the Sun can be considered to be an example of uniform circular motion. Consider a planet of mass m moving about the Sun in a circular orbit of radius r. Suppose the angular velocity of the planet is ω and the mass of the Sun is m_S. The centripetal force needed to keep the planet in orbit is provided by the gravitational attraction between the Sun and the planet.

The orbital period of the planet, $T = \frac{2\pi}{\omega}$ giving $\omega = \frac{2\pi}{T}$

Since gravity provides the centripetal force between the planet and the Sun: $F = mr\omega^2 = \frac{Gm_S m}{r^2}$

Substituting $\omega = \frac{2\pi}{T}$ gives: $\frac{4\pi^2 rm}{T^2} = \frac{Gm_S m}{r^2}$

Rearranging this gives: $T^2 = \frac{4\pi^2 r^3}{Gm_S}$ or $\frac{T^2}{r^3} = \frac{4\pi^2}{Gm_S}$

Since G, m_S and π are constant regardless of the planet being considered, it follows that $\frac{T^2}{r^3}$ = a constant.

Note carefully:
- This relationship between T^2 and r^3 appears in any situation in which one body orbits another.
- In the case of the Earth in orbit around the Sun the mass in the equation is m_S, the mass of the Sun.
- For the Moon in orbit around the Earth the mass in the equation is therefore m_E, the mass of the Earth.

Satellites Orbiting the Earth

For a satellite of mass m, in a circular orbit around the Earth in the equatorial plane and at a distance r from the Earth's centre the gravitational attraction between the satellite and the Earth provides the centripetal force. The relationship between orbital period and orbital radius derived above can be used. Since the satellite is in orbit around the Earth the mass of the Earth m_E is used.

Worked Example

Calculate the orbital period of a satellite in a close orbit of 200 km above the Earth's surface. Use the following data: $m_E = 6.0 \times 10^{24}$ kg, $R_E = 6400$ km.

The radius of the orbit r = 200 km plus the radius of the Earth, R_E = 6400 km, ie:

$r = 2 \times 10^5 + 6.4 \times 10^6$ m $= 6.6 \times 10^6$ m

We know that $G = 6.67 \times 10^{-11}$ N m^2 kg^{-2}

So using $T^2 = \frac{4\pi^2 r^3}{Gm_S} = \frac{4\pi^2 (6.67 \times 10^{-11})^3}{6.67 \times 10^{-11} \times 6.0 \times 10^{24}} = 28367893$ s^2

Hence T = 5326 s or approximately 89 minutes.

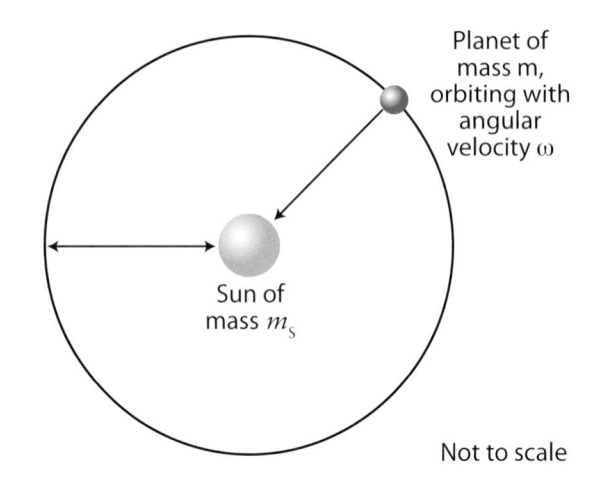

5.2 GRAVITATIONAL FIELDS

Geostationary satellites

A geostationary satellite is one which:
- orbits the Earth directly above the equator,
- orbits in the same direction as the rotation of the Earth, and
- has an orbital period of exactly 24 hours.

These properties mean that the satellite's angular velocity is the same as that of the Earth itself as it rotates on its polar axis. To an observer on the Earth's surface such a satellite appears to be stationary in the sky – hence the name 'geostationary'. This is a very useful property for satellites which are intended to serve a particular part of the Earth's surface, for example communication satellites transmitting TV signals.

Worked Example

(a) State what is meant by a geostationary satellite.
(b) State the period of a geostationary satellite.
(c) State the direction in which a geostationary satellite orbits the Earth.
(d) Comment on the orbital path of a geostationary satellite.
(e) Calculate the radius of the orbit, the height of the satellite above the Earth's surface and the orbital speed.
(f) Comment on your answers to (e).
Use the following data: $m_E = 6.0 \times 10^{24}$ kg, $R_E = 6400$ km.

(a) A geostationary satellite is one that appears to be at rest when viewed by an observer on Earth.
(b) 24 hours.
(c) It orbits the Earth in the same direction as the Earth spins on its axis, ie anticlockwise when viewed from the North pole.
(d) It orbits in a circular path which contains the equatorial plane.
(e) 24 hours = 86 400 s

Using $T^2 = \dfrac{4\pi^2 r^3}{Gm_S}$ gives: $86400^2 = \dfrac{4\pi^2 \times r^3}{6.67 \times 10^{-11} \times 6.0 \times 10^{24}}$ which gives $r = 42311739$ m (or approximately 42 300 km).

Height above Earth's surface = $r - R_E$ = 42 300 km – 6400 km = 35 900 km (approximately).

Orbital speed = $r\omega = 42.3 \times 10^6 \times \dfrac{2\pi}{T} = (42.3 \times 10^6 \times 2 \times \pi) \div 86\,400 = 3076$ m s^{-1}.

(f) We can conclude from the above figures that every satellite in a geostationary orbit travels around the Earth (i) with a fixed period (24 hours) (ii) in a specific direction, the same as the direction of Earth spin on its axis (iii) in a fixed plane, containing the equator (iv) at a fixed height above the surface – approximately 36 000 km – and (v) at a fixed speed of approximately 3080 m s^{-1}.

Exercise 5.2

1. Some data relating to the Earth and planet Jupiter are listed below.

	Mean orbital radius / m	Orbital period / years
Earth	1.49×10^{11}	1.00
Jupiter	7.79×10^{11}	11.9

   (a) Show that the above data are consistent with the mathematical form of Kepler's third law.
   (b) Use the data given for the Earth to calculate the mass of the Sun.
   (c) Assuming the average gravitational force between Jupiter and the Sun is 4.2×10^{23} N, calculate the mass of Jupiter.

2. Two masses, A and B, each of 10 kg are placed 6 m apart as shown below. Find the magnitude and direction of the gravitational field strength at a point P, 5 m from each mass.

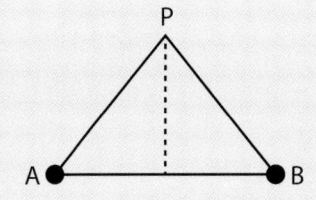

3. The planet Saturn has a radius of 60 475 km, a mass of 5.68×10^{26} kg and it rotates on its axis once every 10.8 hours. Saturn orbits the Sun once every 29.4 years at a mean distance of 1.43 Tm. A satellite is to be put into a geostationary orbit about Saturn. From the information above select the appropriate data and use it to calculate the height of a geostationary satellite above the surface of the planet Saturn.

4. Assume the Earth is a uniform sphere. For the purposes of the gravitational force beyond the surface, a uniform sphere behaves as a point mass with all the mass concentrated at the centre. 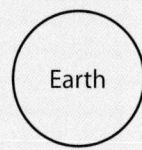
   (a) Copy the diagram above and on it sketch the gravitational field near the surface of the Earth.
   (b) The value of g near the Earth's surface is 9.81 N kg^{-1}. Given that the Earth has a radius of 6400 km, calculate the mass and mean density of the planet. Assume the volume of a sphere of radius r is $\tfrac{4}{3}\pi r^3$.

5. Two point masses, A and B, lie on a straight line. A has a mass of 100 kg and B has a mass of 16 kg. Due to the presence of B there is a neutral point at C which is 0.5 m from A on the line joining A and B. Calculate the distance from A to B.

5.3 Electric Fields

Students should be able to:

5.3.1 Define electric field strength; and

5.3.2 Recall and use the equation $E = \dfrac{F}{q}$;

5.3.3 State Coulomb's law for the force between point charges;

5.3.4 Recall and use the equation for the force between two point charges $F = \dfrac{q_1 q_2}{4\pi\varepsilon_0 r^2} = \dfrac{k q_1 q_2}{r^2}$ where $k = \dfrac{1}{4\pi\varepsilon_0}$ and ε_0 is the permittivity of a vacuum;

5.3.5 Recall and use the equation for the electric field strength due to a point charge, $E = \dfrac{q}{4\pi\varepsilon_0 r^2} = \dfrac{kq}{r^2}$;

5.3.6 Recall that for a uniform electric field, the field strength is constant, and recall and use the equation $E = \dfrac{V}{d}$

5.3.7 Recognise similarities and differences in gravitational and electric fields.

Electric Charges and Coulomb's Law

The diagram on the right shows the electric fields created by two electric charges q_1 and q_2. Each experiences a force since they are within the electric field created by the other charge. Unlike the force between point masses, the force between electrical charges may be attractive or repulsive. The force is **attractive** when the charges have **different** signs. The force is **repulsive** when the charges have the **same** sign.

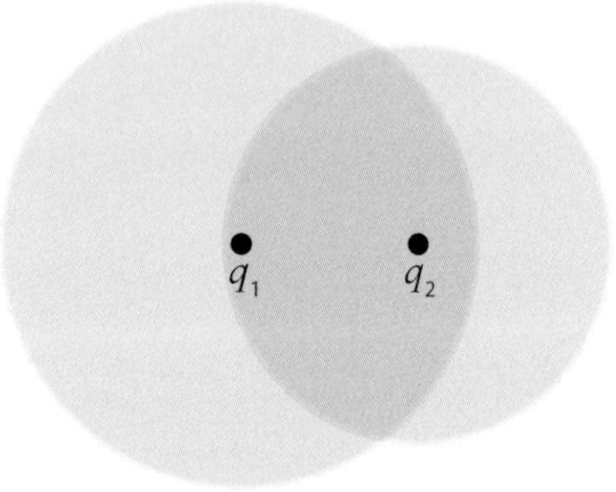

Coulomb's law states that **between every two point charges there exists an electrical force which is directly proportional to the charge of each and is inversely proportional to the square of their separation**. This is written mathematically as:

$F = \dfrac{k q_1 q_2}{r^2}$ where the constant $k = \dfrac{1}{4\pi\varepsilon_0}$

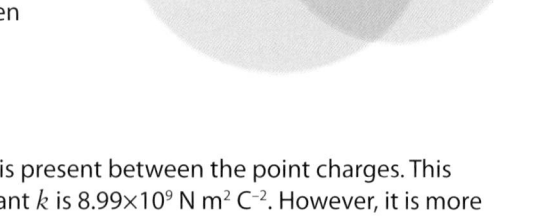

The value of the constant, k, depends on the nature of the material which is present between the point charges. This constant can be found experimentally. For charges in a vacuum, the constant k is 8.99×10^9 N m^2 C^{-2}. However, it is more convenient to use another constant, ε, known as the **permittivity** of the material. For charges in a vacuum we use the value ε_0, known as the **permittivity of free space** (vacuum) which has the experimental value of 8.85×10^{-12} C^2 N^{-1} m^{-2}.

Electric Field Strength

The electric field strength, E, at a point in an electric field is **the force on a charge of 1 C placed at that point**, ie:

$E = \dfrac{F}{q}$ or $F = Eq$ where E = electric field strength (N C^{-1})
F = force (N)
q = charge (C)

Electric field strength is a vector and it has the units N C^{-1}. As you will see later, an alternative unit is V m^{-1}. The direction of the electric field at a given point is the direction in which a positive electric charge would experience a force when placed at that point. Like gravitational fields, electric fields can be illustrated with field lines. Around an isolated positive charge the field lines are radially outwards, while around an isolated negative charge the field lines are radially inwards. Around an isolated positive charge the field lines are radially outwards, while around an isolated negative charge the field lines are radially inwards. The diagrams below show some common electric field patterns. The two on the left show the field around an isolated charge. The two on the right show the field around two adjacent charges.

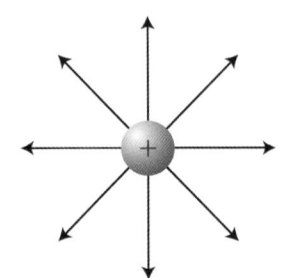

Isolated positive charge 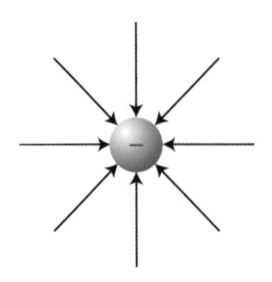   Isolated negative charge 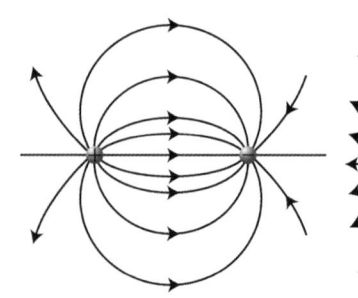   Equal positive and negative charges   Two equal positive charges

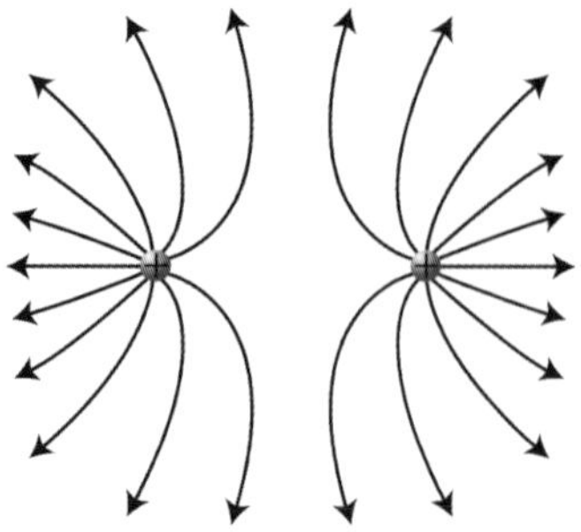

5.3 ELECTRIC FIELDS

Comparison with gravitational fields
The table below summarises the differences and similarities between gravitational and electric fields.

	Gravitational Field	**Electric Field**
Differences	• Acts on masses. • Always produces an attractive force on a mass. • It is impossible to shield an object from a gravitational field.	• Acts on charges. • Can produce both attractive and repulsive forces because there are two types of charge (positive and negative). • Shielding is possible with a suitable material.
Similarities	• Field around a point mass decreases according to an inverse square law (falls off as $1/r^2$). • Field is of infinite range.	• Field around a point charge decreases according to an inverse square law (falls off as $1/r^2$). • Field is of infinite range.

Uniform Electric Field

A uniform electric field has the same field strength throughout. Such a field can be created using a pair of parallel metal plates, a distance d apart, with a constant potential difference V between them, as shown in the diagram on the right.

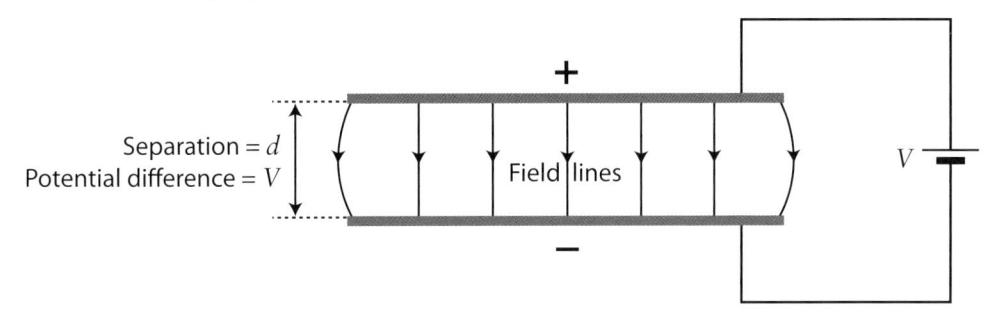

The electric field strength for this uniform field is given by:

$E = -\dfrac{V}{d}$ where E = electric field strength (N C^{-1} or V m^{-1})
V = electrical potential (V)
d = distance between the plates (m)

The minus sign indicates that the electric field strength, E, is in the direction of **decreasing** potential. However it is often omitted, for example when we are only interested in the magnitude of the field. The graph on the right shows how the electric potential varies with distance from the upper (positive) metal plate to the lower (negative) one.

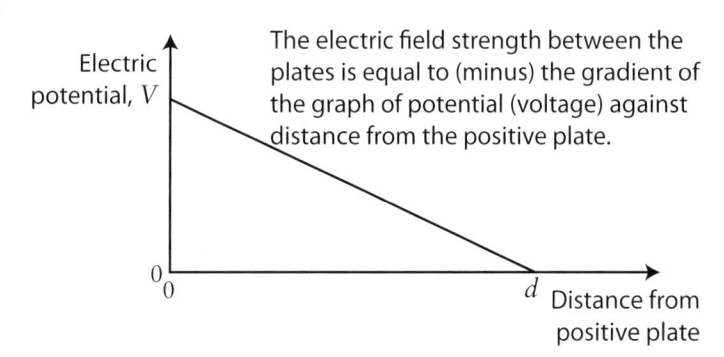

The electric field strength between the plates is equal to (minus) the gradient of the graph of potential (voltage) against distance from the positive plate.

Worked Example

(a) Define electric field strength.

(b) (i) In one model of hydrogen, an electron in its ground state can be considered to orbit the nucleus with a radius of 5.29×10^{-13} m. The nucleus of hydrogen consists of a single proton which can be taken to be a point charge. Calculate the electric field strength due to the proton at this radius.
(ii) Calculate the magnitude of the force between the proton and electron. Assume that the electron is in its ground state.
(iii) State whether the force is attractive or repulsive. Explain your answer.

(a) The force acting on a charge of +1 coulomb at a point in an electric field.

(b) (i) The electric field strength is given by $F = \dfrac{kq_1 q_2}{r^2}$ which, since there is only one charge, becomes $F = \dfrac{kq}{r^2}$

The charge q is the charge on the proton = 1.6×10^{-19} C (from data sheet)

Substitution of values gives $E = \dfrac{8.99 \times 10^9 \times 1.6 \times 10^{-19}}{(5.29 \times 10^{-13})^2}$

$E = 5.14 \times 10^{15}$ N C^{-1}.

(ii) The electric force = electric field strength × the charge
= 5.14×10^{15} × 1.6×10^{-19}
= 8.22×10^{-4} N

(iii) The force is attractive, since the proton is positive and the electron is negative, and unlike charges attract.

Exercise 5.3

1. (a) What is meant by the term "electric field strength"?
   (b) The diagram shows two parallel metal plates which are 15 cm apart in a vacuum. There is a constant potential difference of 75 V between the plates. The direction of the electric field is from the lower plate to the upper plate. 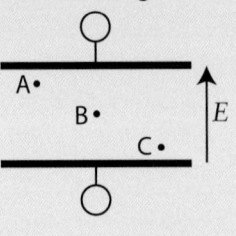

   (i) Copy the diagram and on it indicate in the circles the polarity of the plates.
   (ii) Draw 5 field lines to show the electric field between the plates.

   Points A, B and C are 1 cm, 8 cm and 13 cm respectively from the upper plate. At each of these three points there is an electric charge of -1.6×10^{-3} C. Ignore all repulsive forces between these point charges.

   (iii) At what point, if any, is the electrical force on the charge due to the field set up by the potential difference between the plates the greatest?
   (iv) Calculate the magnitude of the electric field between the plates.
   (v) Calculate the magnitude of the force on the charge at point A and state its direction.

2. Two charged masses, each 5000 kg, are 5 km apart in free space. The magnitude and sign of the charge on each, Q, is the same.
   (a) Find the value of Q if the attractive gravitational force is in equilibrium with the repulsive electrical force.
   (b) A student claims that since the masses are the same, the charges are the same and the gravitational and electrical field strengths both fall off as $1/r^2$, the location of the gravitational and electrical neutral points between the masses must be co-incident. Is the student correct?

3. Two charged polystyrene balls, each of mass 1 g, hang at the ends of identical threads of negligible mass and length 1 m as shown on the right. The balls each carry a charge $+q$. Electrostatic repulsive forces push the balls apart so that at equilibrium they are separated by 20 cm. 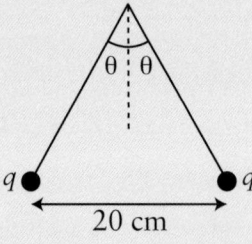
   (a) Calculate the tension in each string when the balls are 20 cm apart.
   (b) Calculate the size of the electrostatic repulsive force between the balls.
   (c) Use your answer to part (b) to calculate the charge q.

4. Triangle ABC is equilateral of side 4.0 m with side AB horizontal and below C. Point charges of -10 mC and $+10$ mC are at points A and B respectively. Assume the medium between the charges is air.
   (a) Find the magnitude and direction of the electric field at point C.
   (b) In what way, if at all, would the answer have changed if each charge was $+10$ mC.

5. State (a) two ways in which gravitational and electrical fields are similar and (b) two ways in which gravitational and electrical fields are different.

5.4 Capacitors

Students should be able to:

5.4.1 Define capacitance;

5.4.2 Recall and use the equation $C = \dfrac{Q}{V}$;

5.4.3 Define the unit of capacitance, the farad;

5.4.4 Recall and use ½QV or its equivalent for calculating the energy of a charged capacitor;

5.4.5 Recall and use the equations for capacitors in series and in parallel;

5.4.6 Perform and describe experiments to demonstrate the charge and discharge of a capacitor;

5.4.7 Confirm the exponential nature of capacitor discharge using V or I discharge curves;

5.4.8 Use the equations $Q = Q_0 e^{-t/CR}$, $V = V_0 e^{-t/CR}$ and $I = I_0 e^{-t/CR}$;

5.4.9 Define time constant and use the equation $\tau = RC$;

5.4.10 Perform and describe an experiment to determine the time constant for R-C circuits;

5.4.11 Apply knowledge and understanding of time constants and stored energy to electronic flash guns and defibrillators;

A capacitor is an electrical component that can store energy in the electric field between a pair of metal plates separated by an insulator, as shown in the diagram. The symbol for a capacitor is also shown.

Charging is the process of storing energy in the capacitor and involves electric charges of equal magnitude, but opposite polarity, building up on each plate.

Capacitance is defined as the **charge stored per volt**. A capacitor's ability to store charge is measured by its **capacitance**, in units of **farads, F**. This can be written in the form of an equation:

$C = \dfrac{Q}{V}$ where C = the capacitance (F)
Q = the charge (C)
V = the potential difference (V)

The farad is a very large unit and microfarads (μF) and picofarads (pF) are more common. 1μF = 1.0×10⁻⁶ F and 1pF = 1.0×10⁻¹² F.

The equation above shows that potential difference is proportional to the amount of charge stored. This can be shown by plotting a graph of Q (vertical axis) and V (horizontal axis). The result is a straight line through the origin, where the gradient is the capacitance. An example of such a graph is shown on the right.

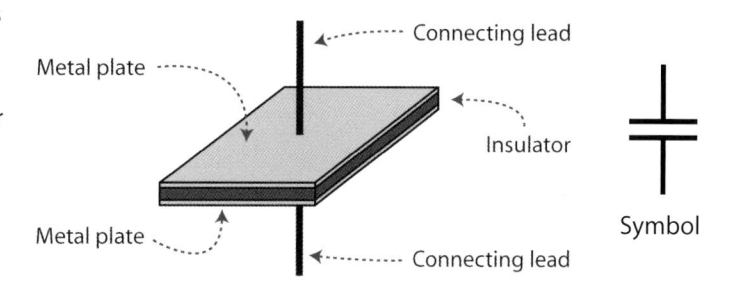

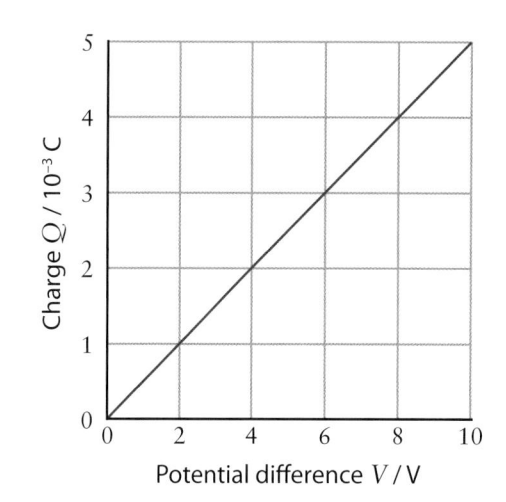

Capacitors in Parallel

In the circuit shown below, the total charge stored is: $Q_1 + Q_2 + Q_3$. The effective, or total, capacitance of this circuit, C, is given by: $C = C_1 + C_2 + C_3$. The total capacitance of any number, N, of capacitors in parallel is the sum of the capacitance of each. This can be written as:

$C = C_1 + C_2 + C_3 + \ldots\ldots C_N$

This shows that we can obtain a larger capacitance by connecting a number of smaller capacitors in parallel.

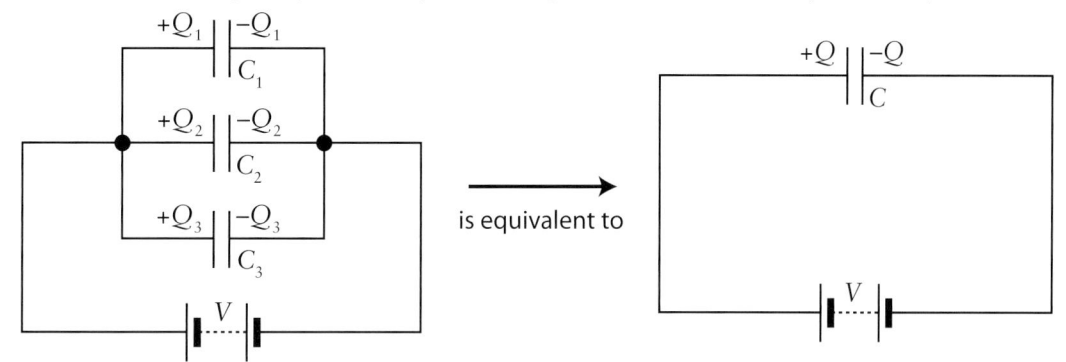

Capacitors in Series

In the circuit shown below, the total charge stored is Q, not $3Q$. The effective, or total, capacitance of this circuit, C, is given by: $\dfrac{1}{C} = \dfrac{1}{C_1} + \dfrac{1}{C_2} + \dfrac{1}{C_3}$. The total capacitance of any number, N, of capacitors in series is given by the equation:

$$\dfrac{1}{C} = \dfrac{1}{C_1} + \dfrac{1}{C_2} + \dfrac{1}{C_3} + \ldots \dfrac{1}{C_N}$$

This shows that we can obtain a smaller capacitance by connecting a number of larger capacitors in series.

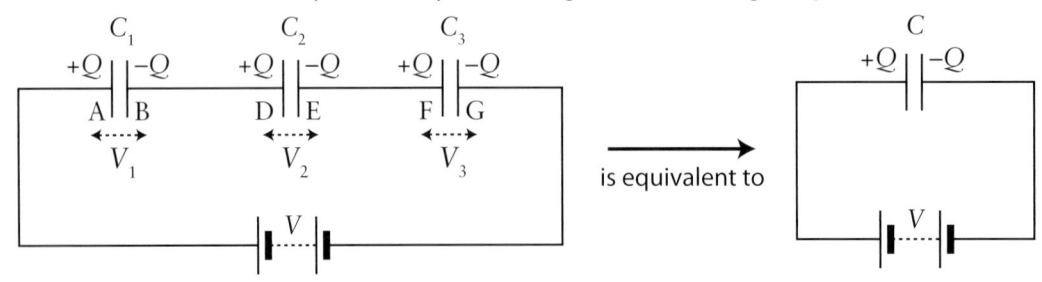

Worked Example

(a) Two capacitors of capacitance 16 µF and 8 µF are connected in parallel. Find their total capacitance.
(b) The two capacitors are now connected in series. Find their new total capacitance.

(a) Total capacitance $C = C_1 + C_2 = 16 + 8 = 24$ µF

(b) Using $\dfrac{1}{C} = \dfrac{1}{C_1} + \dfrac{1}{C_2} = \dfrac{1}{16} + \dfrac{1}{8} = \dfrac{3}{16}$. Therefore $C = \dfrac{16}{3} = 5.33$ µF (to 3 significant figures)

Energy of a Charged Capacitor

The energy, E, stored by a capacitor is given by: $E = \tfrac{1}{2}QV$ where E = energy stored (J)
Q = charge (C)
V = potential difference (V)

Using the relationship $C = \dfrac{Q}{V}$ alternative equations can be derived for the energy: $E = \tfrac{1}{2}CV^2$ and $E = \dfrac{Q^2}{2C}$.

The relationship can be verified by experiment. A capacitor is first charged using a known voltage, and is then discharged through a joule meter to measure the total energy stored. The graphs below show the results for a capacitor of capacitance 6800 mF. The left graph shows energy stored, E, plotted against the potential difference, V, producing a curve. The right graph shows energy stored, E, plotted against V^2, producing a straight line through the origin, thus verifying that $E = \tfrac{1}{2}CV^2$.

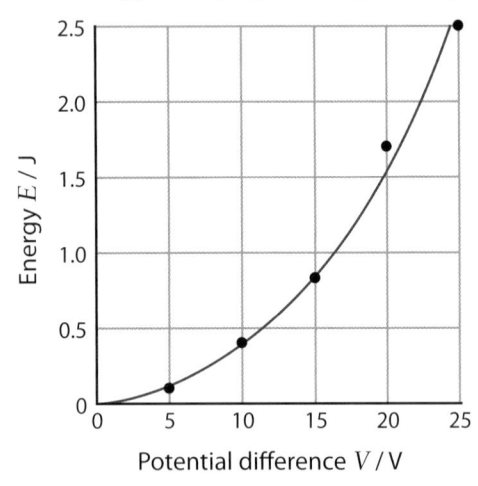

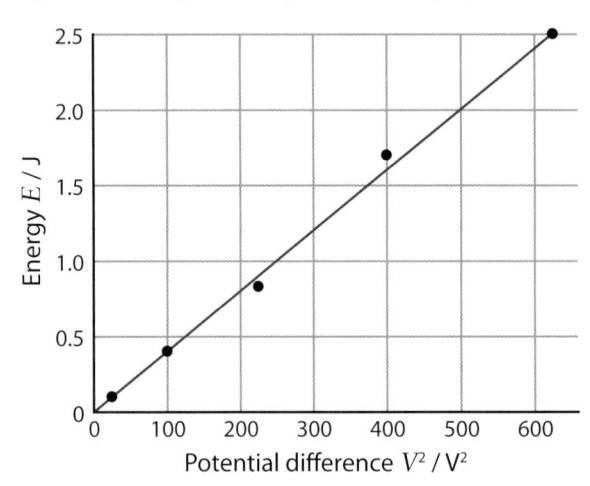

Note: The equation $E = \tfrac{1}{2}QV$ might suggest that the energy, E is proportional to V. This is not the case. Remember that as V increases so does Q. The equation $E = \tfrac{1}{2}CV^2$ shows that the energy is actually proportional to V^2, since the capacitance C is constant for a given capacitor. The graphs above demonstrate this to be true.

Charging and Discharging Capacitors

Charging a Capacitor

The circuit on the next page shows a capacitor connected in series with a battery, a switch and a resistor. When the switch is closed the capacitor charges. Electrons flow onto one plate, which causes electrons to be repelled from the other plate. This movement of electrons constitutes a current. The charging current decreases and eventually stops flowing once the potential difference across the capacitor is exactly equal to the e.m.f. of the battery.

The rate at which the charging current decreases and the rate at which the potential differences across the capacitor rises depends on the **capacitance** of the capacitor and the **resistance** of the circuit.

The graph on the right shows how the charging current varies with time. Notice how the charging current **decreases** with time. Initially there are no electrons on the negative plate, so the rate of flow of electrons (the current) is therefore large. It decreases because it becomes more difficult to push electrons onto the plate because of the repulsion of the electrons already there. The lower the resistance of the circuit, the more rapidly the capacitor charges, and hence the charging current falls more quickly.

As the capacitor stores an increasing amount of charge so the potential difference across it also increases, as shown in the second graph. Eventually this potential difference reaches the e.m.f. of the charging battery and at this point the flow of electrons ceases.

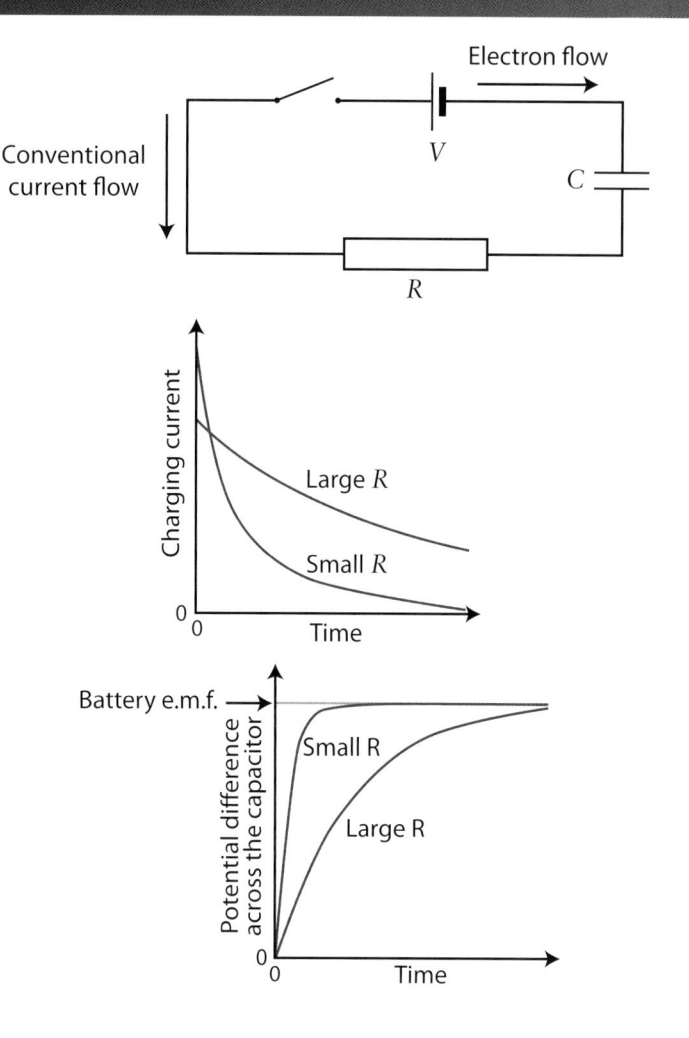

Discharging a Capacitor

The R-C circuit shown on the right can be used to investigate the discharge of a capacitor. When the switch is moved to position A the capacitor is charged from the battery. The capacitor becomes fully charged very quickly since the resistance of the charging circuit is very small. Moving the switch to position B will start the capacitor discharging through the resistor R. The reading on the microammeter will start high and gradually fall. Values of the current should be recorded at 10 or 20 second intervals.

The results from such an experiment are shown in the graph below. Note that as the capacitor discharges, the current flowing decreases exponentially with time.

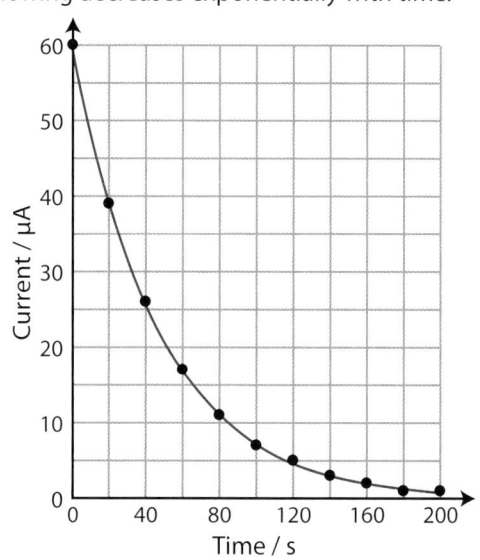

The Time Constant

The product of the capacitance and the resistance of a circuit is known as the **time constant**, τ, of the circuit. This can be expressed in the form of an equation:

$\tau = RC$ where τ = time constant (s)
R = resistance of the circuit (Ω)
C = capacitance (F)

We have seen that the current flowing in a discharging capacitor decreases exponentially. The equation that describes this decrease is as follows:

$I = I_0 e^{\frac{-t}{\tau}}$ where I_0 = the initial current (A)
I = current at time t (A)
t = time (s)
τ = time constant (s)

Since $\tau = RC$, this equation is also sometimes written as: $I = I_0 e^{\frac{-t}{CR}}$

PHYSICS FOR CCEA A2 REVISION GUIDE, 2ND EDITION

Significance of the Time Constant

The time constant, τ, is the time take for the current to fall to $1/e$ of its initial value. The value of e is approximately 2.7183, so the current after one time constant, τ, is given by:

$$I = \frac{I_0}{2.7183} \approx 0.37 \times I_0$$

After a time equal to 2τ the current will fall to 0.37 of 0.37, i.e. to approximately 0.14 of its initial value. This is an exponential decline and is shown in the graph on the right.

The product of capacitance and resistance (the time constant) has the unit of time and you can show this as follows:

$$C = \frac{Q}{V} \text{ and } R = \frac{V}{I} \text{ so therefore } CR = \frac{Q}{V} \times \frac{V}{I} = \frac{Q}{I}$$

However, current = charge × time, so $Q = It$

Therefore $CR = \frac{It}{I} = t$ (time)

Note that for a discharging capacitor, graphs of **voltage** against time and **charge** against time have the same shape as the graphs of current against time that are shown in the preceding discussion.

Worked Example

A R-C circuit has a capacitance of 470 μF and resistance of 100 kΩ. Calculate the time constant.
$C = 4.7 \times 10^{-4}$ F and $R = 1 \times 10^5$ Ω
Using $\tau = CR = 4.7 \times 10^{-4} \times 1 \times 10^5 = 47$ s

---

Note that the charge, Q, stored by the capacitor and the potential difference, V, across the capacitor also decrease in an exponential way. The equations that describe the decrease of each are as follows:

$Q = Q_0 e^{\frac{-t}{\tau}}$ where Q_0 = the initial charge (C)
Q = charge at time t (C)
t = time (s)
τ = time constant (s)

$V = V_0 e^{\frac{-t}{\tau}}$ where: V_0 = the initial potential difference (V)
V = potential difference at time t (V)
t = time (s)
τ = time constant (s)

Measuring the Time Constant for a R-C Circuit

A circuit for investigating the discharge of a capacitor has already been discussed. Values of current against time can be plotted directly and an exponential decay curve plotted. The time constant can be determined from the graph by measuring the time taken for the current to fall to $1/e$ or 0.37 of its initial value. However, drawing exponential curves by hand is difficult and a better method is to obtain a straight line graph.

Plotting a graph of **ln I** on the y-axis and **time t** on the x-axis gives a straight line as shown on the right. The gradient of the line will give the value of the time constant τ. Since $\tau = CR$, if either the capacitance or the resistance is known, the other can be found.

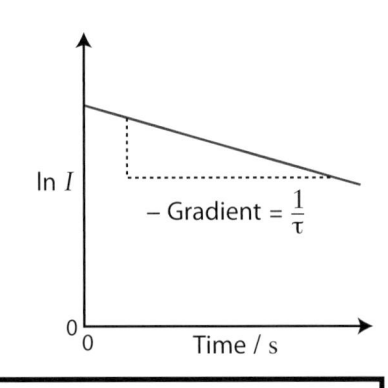

Worked Example

The diagram shows a circuit which can be used to charge a capacitor C of capacitance 6 μF. The switch A is closed at time t = 0.
(a) Sketch a graph to show how the voltage V, recorded by the voltmeter varies with time t during the charging of the capacitor. Label each axis and label the final value for V on the y-axis.
(b) Explain how the movement of charge carriers in the circuit, after the switch is closed, can explain the shape of the graph you have drawn.
(c) The switch is now moved to position B and the capacitor discharges through resistor R. After 48 seconds, the voltage has fallen from 12.0 V to 1.64 V. Use these data to calculate the resistance of R.

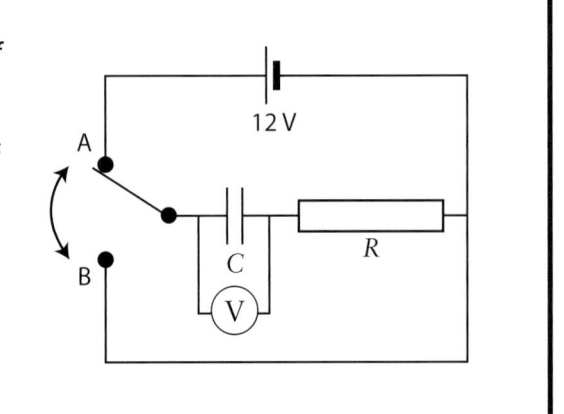

(a) The voltage increases from zero. The rate of increase decreases with time. See graph on the right.
(b) Initially, there is a rapid flow of charge electrons because the capacitor is uncharged. As the number of electrons increases the force repelling further electrons increases. This slows the increase in the voltage and charge being stored on the plates of the capacitor.
(c) *You can use either of two methods.*
Firstly, divide voltage at time t = 48 s by the initial voltage:
$1.64 \div 12 = 0.137$. This equals 0.37×0.37, which means that 48 s is two time constants. Therefore $2\tau = 48$ s, and hence $\tau = 24$ s.

Secondly, you can use $V = V_0 e^{\frac{-t}{\tau}}$

$1.67 = 12\, e^{\frac{-48}{\tau}}$ so $0.137 = e^{\frac{-48}{\tau}}$

Taking natural logs of both sides gives:

$-1.99 = \frac{-48}{\tau}$ so $\tau = \frac{48}{1.99} = 24$ s

$\tau = CR$, so $R = \frac{\tau}{C} = 24 \div 6 \times 10^{-6} = 4 \times 10^6\ \Omega = 4\ M\Omega$

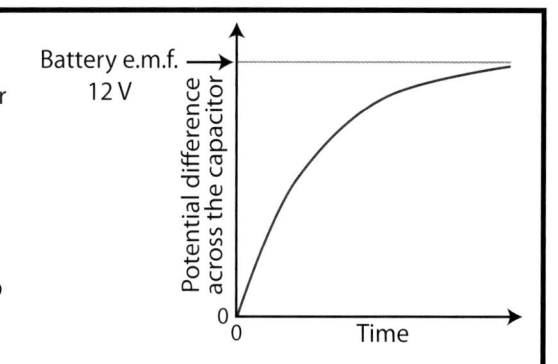

Uses of Capacitors

Defibrillators deliver a carefully controlled shock to a heart attack victim, whose heart muscles are twitching in an uncoordinated fashion known as ventricular fibrillation. The controlled shock is designed to stop the fibrillation and start the normal heart rhythm again. A defibrillator needs to transfer a precise amount of energy to a patient. The best way to do this is to use a capacitor. The capacitor stores electric charge on its plates. The energy that it stores is in the form of the electric field that is created between its plates. This energy is delivered as a pulse lasting just a few milliseconds.

Electronic flash guns use an electric discharge in a suitable gas such as xenon to produce an intense flash of light that lasts a short time. In the case of a flash gun used with a camera, it must be operated from a small battery, say, 6 V. To achieve the flash, electronic circuitry has to be used to generate a high voltage, ie several hundred volts. When the shutter on the camera is pressed the charged capacitor is rapidly discharged through the gas-filled flash tube so producing the intense flash of light. Today flash guns used in photographic studios might use capacitors ranging from about 4 mF to about 20 mF at about 300 V. The output energy might therefore range from 180 J to about 900 J. The flash may last for only 5 ms or so, so the output power might range from 36 kW to 180 kW.

Exercise 5.4

1. The aluminium sphere of a Van de Graaf generator is charged to a potential of 120 kV. The charge stored on the sphere is 3 µC. Calculate:
   (a) the capacitance of the sphere,
   (b) the electrical energy stored.

2. You are provided with a large number of 3 µF capacitors. Using the minimum possible number of capacitors, how might you arrange some of them to produce a capacitor of capacitance (a) 1 µF (b) 2 µF (c) 4 µF and (d) 5 µF?

3. A 100 µF capacitor is connected across a 12 V DC electrical supply.
   (a) Calculate (i) the charge and (ii) the energy stored in the capacitor.

   The charged capacitor is now disconnected from the battery. It is then connected in parallel with an uncharged 300 µF capacitor.
   (b) (i) Calculate the final voltage across the parallel combination of capacitors.
      (ii) Calculate the electrical energy now stored in the parallel combination of capacitors.
      (iii) Explain why the energy found in (b)(ii) is less than that found in (a)(ii).

4. The diagram below shows a circuit used to investigate the charging and discharging of a capacitor.

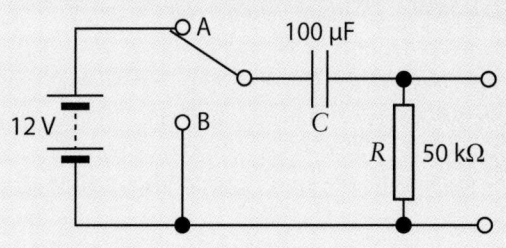

   (a) The switch is closed at A and the capacitor starts to charge.
      (i) Calculate the time constant for the charging circuit.
      (ii) Calculate the maximum charge which can be stored on the capacitor in this circuit.
      (iii) Sketch a graph to show how the charge stored on the capacitor changes with time.
      (iv) Explain in terms of electrons why there comes a point when no additional charge can be stored in the capacitor.
      (v) The charge stored on the capacitor at any time, t, is given by the equation:

   $$Q = Q_0 e^{\frac{-t}{\tau}}$$

where Q_0 is the maximum charge stored in the capacitor and τ is the circuit time constant. Calculate the time taken for the charge stored to reach half of its maximum value.

(b) When the voltage across the capacitor is 12.0 V, the switch is closed at B and the capacitor starts to discharge.
   (i) Calculate the current which flows through the resistor at the moment the switch is closed.
   (ii) Calculate the charge stored in the capacitor 4 seconds after the switch is closed.
   (iii) At what time after closure of the switch is the voltage across the resistor equal to 3 V?

5.5 Magnetic Fields

Students should be able to:

5.5.1 Describe the shape and direction of the magnetic field produced by the current in a coil of wire and a long straight wire;

5.5.2 Demonstrate an understanding that there is a force on a current-carrying conductor in a perpendicular magnetic field and be able to predict the direction of the force;

5.5.3 Demonstrate an understanding that the forces produced on a current-carrying coil in a magnetic field is the principle behind the electric motor;

5.5.4 Recall and use the equation $F = BIl$;

5.5.5 Define magnetic flux density;

5.5.6 Demonstrate an understanding of the concepts of magnetic flux and magnetic flux linkage;

5.5.7 Recall and use the equations for magnetic flux, $\Phi = BA$, and magnetic flux linkage, $N\Phi = NBA$;

5.5.8 State, use and demonstrate experimentally Faraday's and Lenz's laws of electromagnetic induction;

5.5.9 Recall and calculate average induced e.m.f. as rate of change of flux linkage with time;

5.5.10 Demonstrate an understanding of the simple a.c. generator and use the equation $E = BAN\omega \sin \omega t$;

5.5.11 Describe how a transformer works;

5.5.12 Recall and use the equation $\dfrac{V_s}{V_p} = \dfrac{N_s}{N_p} = \dfrac{I_p}{I_s}$ for transformers;

5.5.13 Explain power losses in transformers and the advantages of high-voltage transmission of electricity.

The space surrounding a magnet where a magnetic force is experienced is called a **magnetic field**. The direction of a magnetic field at a point is taken as the direction of the force that acts on a north pole placed at that point. The shape of a magnetic field can be represented by magnetic field lines, also known as magnetic flux lines. Since a north pole is repelled from another north pole and attracted by a south pole the **direction of a magnetic field is from north to south**.

Magnetic Field Produced by an Electric Current in a Long Straight Wire

If iron filings are sprinkled on a card, and a DC current from a battery is then passed through the card, the filings will show the magnetic field as a series of circles as shown in the left diagram below. The field direction can be predicted by the **right hand grip rule**: Grasp the conductor with the right hand with the thumb pointing in the direction of the current. The direction of twisting of the fingers gives the direction of the lines of the magnetic field as shown on the right below.

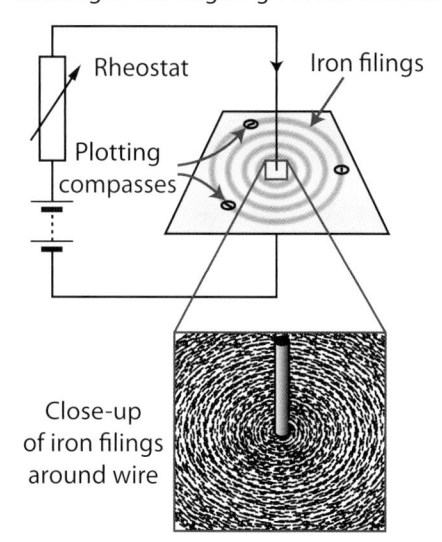

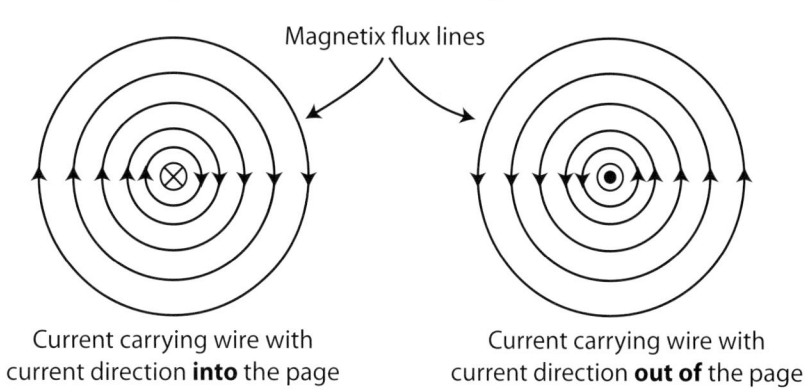

Note that (1) as we move further from the conducting wire the field lines get further apart, showing that the strength of the field decreases with increasing distance from the wire, and (2) the strength of the magnetic field produced at any point is directly proportional to the current in the wire.

Magnetic Field Produced by an Electric Current in a Coil of Wire

The diagram on the right shows the shape of the magnetic field around a single loop of wire when it carries an electric current. At A, the current is flowing **out of** the plane of the paper and the magnetic field lines there turn anti-clockwise. At B, the current is flowing **into** the plane of the paper and the magnetic field lines there point clockwise. In the middle, the fields from each part of the loop combine to produce a magnetic field pointing from bottom to top.

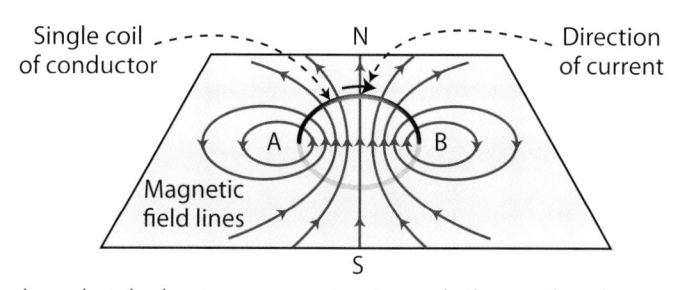

Note the field lines at the top diverging away from the north pole and at the bottom converging towards the south pole.

PHYSICS FOR CCEA A2 REVISION GUIDE, 2ND EDITION

Coil With Many Turns

The diagram on the right shows the shape of the magnetic field caused by the current in a long, cylindrical coil or **solenoid**. Within the solenoid the field lines are parallel and equally spaced (so the field there is uniform). Outside the solenoid the field lines loop, in a way which is similar to that around a bar magnet. As they leave the solenoid the field lines diverge as the strength of the field at that point decreases. The direction of the field can be found by using the right hand grip rule.

Note that the strength of the magnetic field produced can be increased by (1) increasing the current in the coil (2) increasing the number of turns in the coil or (3) placing a rod of soft iron in the coil with its axis perpendicular to the plane of the coil.

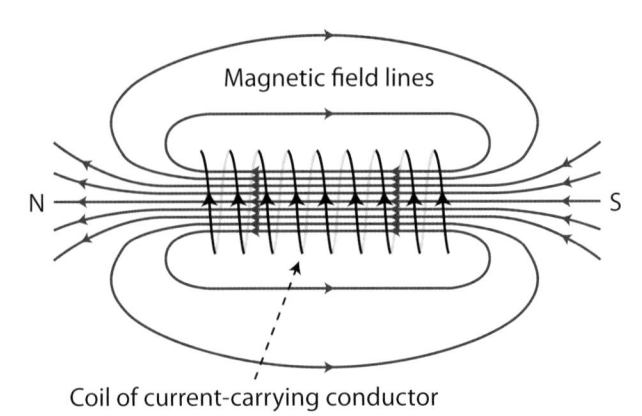

Force on a Current-Carrying Conductor in a Perpendicular Magnetic Field

When a current flows through a conductor which is placed in a magnetic field, the conductor experiences a **force**. The diagram on the right shows a flexible wire in a magnetic field. When a current is passed along the flexible wire, the wire moves up. The force acts at right angles to both the current and the magnetic field direction.

The direction of the force (movement) of the wire is obtained from Fleming's left hand rule, which is shown on the far right.

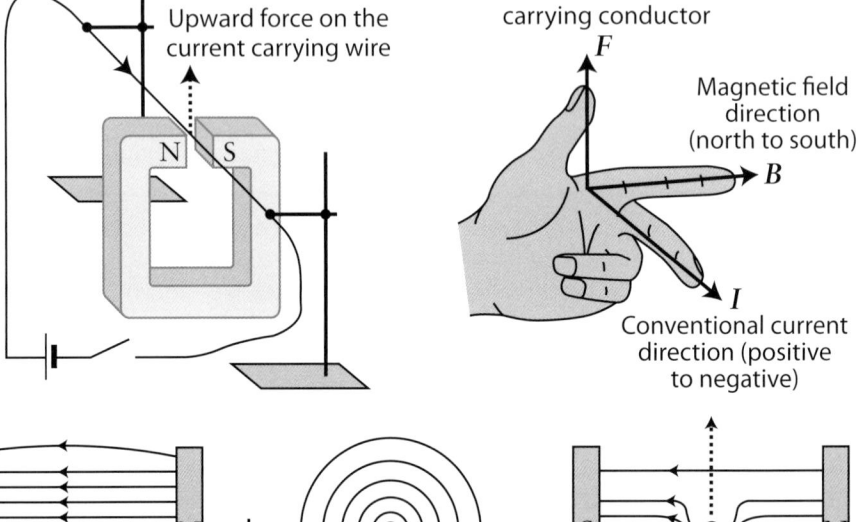

This force arises because of the interaction of the two magnetic fields: the uniform field of the permanent magnet and the field due to the current in the wire. The magnetic field lines of force are **vectors** and the field lines due to two fields have to be combined vectorially, as shown on the right.

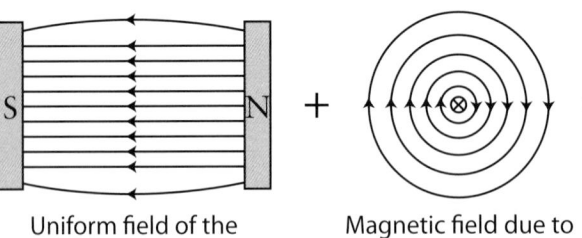

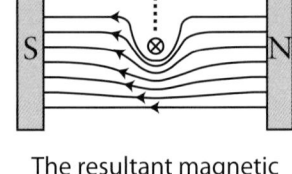

Magnetic Flux Density

In the case when the magnetic field and the current are at right angles the force is given by:

$F = BIl$ where: B = magnetic flux density (measured in teslas, symbol T)
I = current in the conductor (A)
l = the length of the conductor in the magnetic field (m)

Definition of the Tesla

Re–arranging the above equation allows us to give a definition of magnetic flux density as: $B = \dfrac{FI}{l}$.

This shows that magnetic flux density is **the force per unit current carrying length**. A 'current carrying length' is the product of current and length. For example, a current of 0.25 A in a wire of length 0.4 m is a current carrying length of $0.25 \times 0.4 = 0.1$ Am. This allows us to define the tesla. If a current of 1 A flowing in a conductor at right angles to a magnetic field causes a force of 1 N to be produced on each metre of conductor within the field, then the strength of the magnetic field is 1 tesla, or 1 T. **Note** that the force, current and magnetic field directions are perpendicular to each other in accordance with Fleming's left hand rule.

> ### Worked Example
> *A straight wire of length 50 cm carries a current of 1.75 A. Calculate the value of the force that acts on this wire when a length of 30 cm of this wire is placed at right angles to a magnetic field of flux density 5.5×10^{-2} T.*
>
> Use $F = BIl$. Only 30 cm of the wire is in the magnetic field, so $l = 0.30$ m. So $F = BIl = 5.5 \times 10^{-2} \times 1.75 \times 0.30 = 0.029$ N

5.5 MAGNETIC FIELDS

Verification of $F = BIl$

The apparatus shown on the next page can be used to investigate how the force acting on a current carrying conductor depends on the current flowing in the conductor and the length of the conductor in the magnetic field. An aluminium rod is clamped horizontally above a sensitive electronic balance. The rod is connected to a variable low voltage supply. An ammeter connected in series with both will allow the current to be measured.

A permanent magnet is placed on the balance and the aluminium rod positioned so that it is located in the centre of the magnetic field. The balance is set to read zero after the magnets have been placed on it. When a current is then passed along the clamped aluminium rod the rod experiences a force due to the interaction of the permanent magnetic field and the magnetic field due to the current in the aluminium rod. In the case shown in the diagram, the force is upwards. Use Fleming's Left Hand Rule to verify this.

The current is varied with a single magnet in place. This ensures that the length of the conductor in the magnetic field remains constant.

A graph of the force (y-axis) against current (x-axis) produces a straight line through the origin, as shown in the first graph on the right.

This is verification that the force on the current carrying conductor in the magnetic field is directly proportional to the **current**.

The current is then fixed and a number of identical magnets are placed side by side. This changes the length of the conductor in the magnetic field. A graph of force (y-axis) against the length of conductor in the magnetic field (x-axis) produces a straight line through the origin, as shown in the second graph.

This is verification that the force on the current carrying conductor in the magnetic field is directly proportional to the **length of the conductor** in the magnetic field.

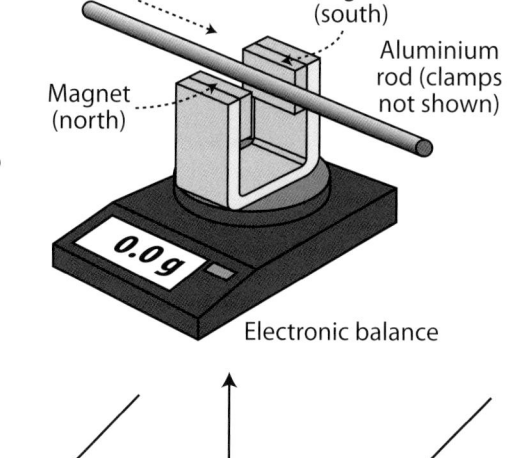

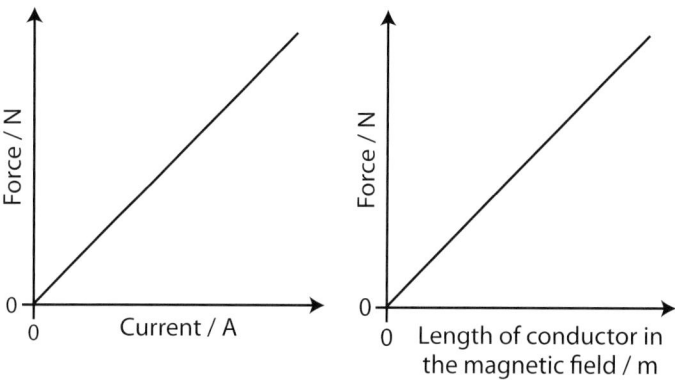

The Principle of the Electric Motor

In the diagram on the right, the rectangular loop of wire lies between the poles of a magnet. The current flows in opposite directions along the two sides of the loop. Application of Fleming's left hand rule shows that one side of the loop is pushed up and the other side is pushed down. This causes a turning effect on the loop. If the number of loops is increased to form a coil, the turning effect is greatly increased. This is the principle involved in electric motors.

The diagram below right shows a simple DC motor. The coil, made of insulated copper wire, is free to rotate on an axle between the poles of a magnet.

When the coil is horizontal, the forces are furthest apart and have their maximum turning effect on the coil. There is no turning force when the plane of the coil is vertical. However, as the coil overshoots the vertical, the commutator changes the direction of the current through it. This means that the forces change direction and so the coil keeps rotating clockwise. You do not need to know the details of the action of the commutator. If either the battery or the poles of the magnet are reversed, the coil will rotate anticlockwise.

The turning effect on the coil can be increased by increasing:
- the current in the coil,
- the number of turns on the coil,
- the strength of the magnetic field,
- the area of the coil.

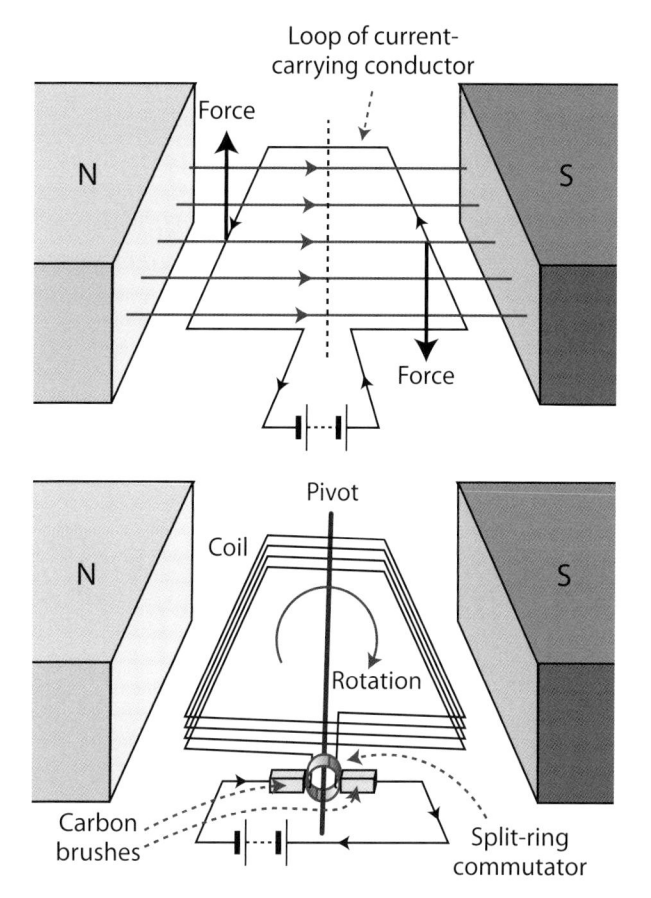

55

Magnetic Flux Φ

Magnetic flux lines (magnetic field lines) show the direction of a magnetic field. Their spacing indicates the strength of the field: the closer the field lines, the stronger the magnetic field. Magnetic flux, Φ, represents the total number of magnetic flux lines that pass at 90° through a given area. Magnetic flux is measured in webers (Wb) and is given by:

$\Phi = BA$ where: Φ = magnetic flux (Wb)
                              B = magnetic flux density, perpendicular to the plane of the coil (T)
                              A = the area (m²)

If the field, B, is at an angle θ to the normal of the area of the coil, A (as shown on the right) then the magnetic flux is given by the product of the area and the component of B which is normal to the plane of the coil, ie:

$\Phi = BA \cos \theta$

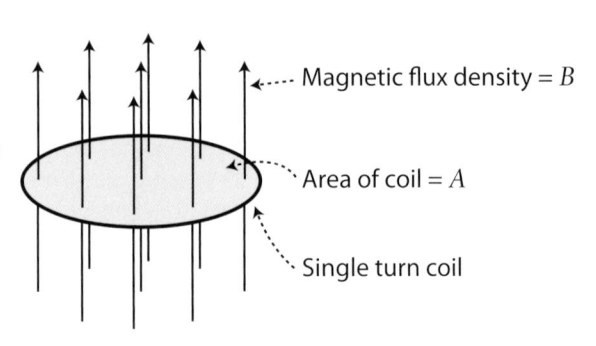

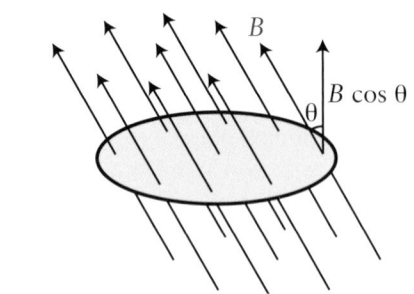

Magnetic Flux Linkage NΦ

Magnetic flux linkage is used when calculating the total magnetic flux passing through or linking a coil of N turns and area of cross section, A. Magnetic flux linkage is also measured in webers (Wb), since N does not have a unit. Magnetic flux linkage is given by:

$N\Phi = BAN$ where: Φ = magnetic flux (Wb)
                              B = magnetic flux density, perpendicular to the plane of the coil (T)
                              A = the area (m²)
                              N = number of turns in the coil

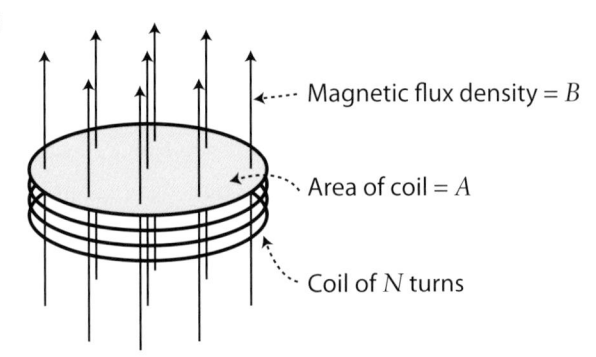

Worked Example

(a) A coil of 200 turns, each of diameter 5 cm, is placed in a uniform magnetic field of flux density 0.5 T. The magnetic field direction is perpendicular to the plane of the coil. Calculate the magnetic flux linkage.
(b) The coil is now rotated so that the magnetic field direction makes an angle of 30° with the normal to the plane of the coil. Calculate the new magnetic flux linkage in the coil.

(a) Use $N\Phi = BAN$. The area $A = \pi r^2 = \pi \times 5^2 = 78.5$ cm² $= 7.85 \times 10^{-3}$ m²
So $N\Phi = BAN = 0.5 \times 7.85 \times 10^{-3} \times 200 = 0.785$ Wb

(b) Only the perpendicular component of the magnetic flux density, $B_\perp$ passes through the coil.
$B_\perp = B \cos 30 = 0.5 \cos 30 = 0.5 \times 0.866 = 0.433$ T
So new $N\Phi = B_\perp AN = 0.433 \times 7.85 \times 10^{-3} \times 200 = 0.680$ Wb

Electromagnetic Induction

An electromotive force (e.m.f.) can be **induced** in a coil of wire by moving a magnet towards or away from the coil, or by moving a wire so that it cuts across the magnetic lines of flux. For an e.m.f. there must be relative motion between the magnet and the conductor, the wire or coil of wire.

The magnitude of the e.m.f. is proportional to:
- the strength of the magnet,
- the number of turns on the coil,
- the speed of the moving magnet.

This can be stated formally as **Faraday's law** of electromagnetic induction:

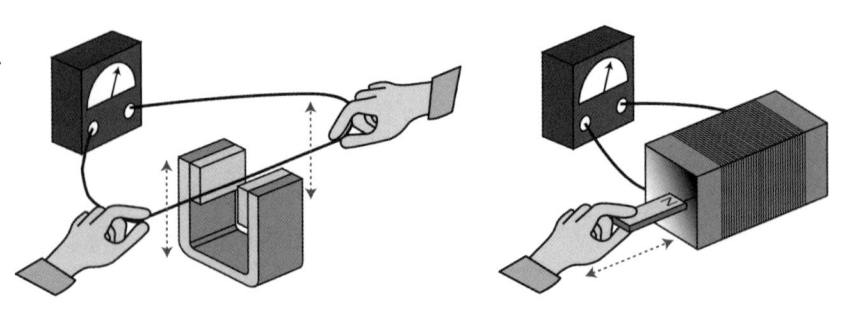

5.5 MAGNETIC FIELDS

The magnitude of the induced e.m.f. is equal to the rate of change of magnetic flux linkage.

The direction of the induced e.m.f. depends on the direction in which the magnet is moving and on the type of magnetic pole nearest the coil. We can demonstrate this using the simple apparatus shown below.

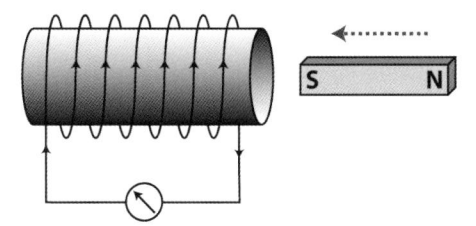

Moving the south pole of the magnet **towards** the coil causes the induced current to flow so that it creates a south magnetic pole in the coil opposing the incoming south pole of the magnet. Work has to be done against this opposing force.

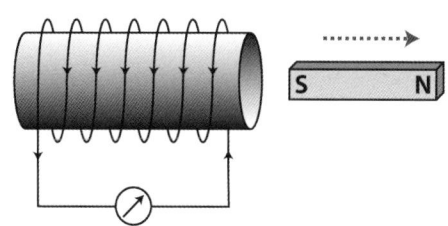

Moving the south pole of the magnet **away from** the coil causes the induced current to reverse direction. It now flows so that it creates a north magnetic pole in the coil attracting the retreating pole of the magnet. Work again has to be done against this opposing force.

These observations can be stated formally as **Lenz's law**:

The direction of the induced current is such that it opposes the change in the magnetic flux that is producing it.

Lenz's law is the principle of conservation of energy in action. The kinetic energy of the moving magnet is converted to electrical energy when work is done against the opposing force.

Calculation of Induced e.m.f.

Faraday's Law states that the size of the induced e.m.f. is equal to the rate of change of the number of magnetic field lines passing through the coil. The average induced e.m.f. can be calculated as follows.

Average induced e.m.f. = rate of change of magnetic flux linkage with time, or as an equation:

$E = -\dfrac{\Delta N\Phi}{\Delta t}$ where E = induced e.m.f (V)
$\Delta N\Phi$ = change in the magnetic flux linkage (Wb)
Δt = time in which the change in the magnetic flux linkage occurs (s)

The minus is a consequence of Lenz's Law.

The AC Generator

The diagram on the right shows the structure of a simple AC (alternating current) generator which can be used to create an electric current from rotational kinetic energy. It consists of a coil of wire that is rotated at a constant angular velocity in a magnetic field. As the coil turns, the magnetic flux linking it changes. This change in magnetic flux linkage results in the induction of an alternating e.m.f. in the coil. The output from the rotating coil is led to the outside by means of carbon brushes that rub against metal slip rings.

Note: In a practical generator there would be a lot more turns on the coil. The coil would be formed around a soft iron core. These two changes produce an alternating output voltage with a much greater peak value.

Coil is turned in a magnetic field

Slip rings and brushes

Output is an alternating e.m.f.

Suppose a coil of N turns and area A is rotating with an angular speed ω in a constant magnetic field of strength, B. Suppose also that at time $t = 0$, the plane of the coil is perpendicular to the field so that the flux linkage with the coil is a maximum. Then at any given time t, $N\Phi = BAN \cos \omega t$

Because of the equation $E = -\dfrac{\Delta N\Phi}{\Delta t}$ we can obtain the e.m.f. by taking the negative of the gradient of the flux linkage graph, ie:

$E = -\dfrac{d(N\Phi)}{dt} = \dfrac{d(BAN \cos \omega t)}{dt}$

From the rules of trigonometry, this results in a sine function:

$E = BAN\omega \sin \omega t$ where E = induced e.m.f (V)
ω = angular velocity (rad s^{-1})
and the other symbols have their usual meanings

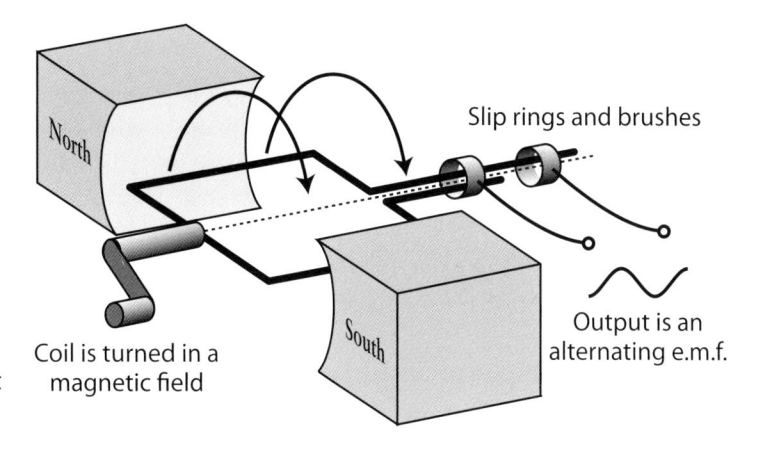

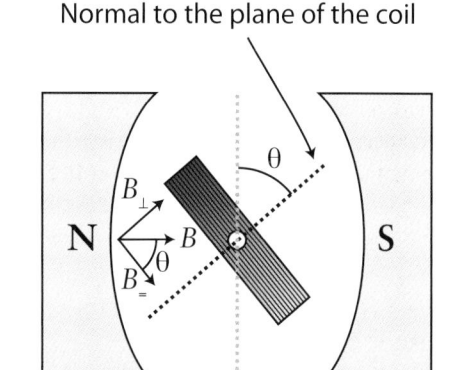

57

Worked Example

An AC generator consists of rectangular coil measuring 30 cm by 20 cm and has 25 turns. It rotates at a uniform rate of 3000 revolutions per minute about an axis parallel to its long side and perpendicular to a uniform magnetic field of flux density 50 mT. Calculate (a) the maximum flux linked with the coil, (b) the frequency of the AC voltage output from the generator and (c) the maximum voltage output from the generator. (d) In what two ways would the output from the generator change if the coil rotated at 6000 rpm?

(a) Maximum flux linkage = NBA = 25 × 0.05 × (0.3 × 0.2) = 0.075 Wb-turns
(b) f = number of revolutions per second = 3000 ÷ 60 = 50 Hz
(c) Maximum value of $E = BAN\omega$ = 0.075 × (2 × π × 50) = 23.6 V
(d) The output voltage would double to 47.2 V and the frequency would double to 100 Hz.

The following example illustrates an important point relating to the **average** induced e.m.f. over a period of time.

Worked Example

A solenoid A is connected to a 50 Hz alternating voltage supply. A second solenoid B is positioned 10 cm from the first where the maximum flux density is 1.6 mT. Calculate the average electromotive force (e.m.f.) induced in solenoid B over a quarter period if it has an area of cross section of 0.0048 m² and contains 200 turns.

The flux density is varying sinusoidally with a peak value of 1.6 mT. Consider the change from 1.6 mT to zero. This happens in a quarter-period and, since the frequency is 50 Hz, the period is 20 ms. A quarter period is therefore 5 ms or 0.005 s. So the average rate of change of flux density = change in flux density ÷ time taken = 0.0016 ÷ 0.005 = 0.32 T s^{-1}.

The average induced e.m.f. = average rate of change of flux linkage
= average rate of change of flux density × area × number of turns = 0.32 × 0.0048 × 200 = 0.307 V.

The Transformer

A transformer is a device that is used to either increase ('step up') or decrease ('step down') the supplied voltage. This is very useful for applications such as power transmission, as shall be discussed later.

The principle of the transformer can be demonstrated using two coils arranged as shown in the diagram on the right. The primary coil is connected to an AC power supply and a switch. The secondary coil is connected to a sensitive ammeter. A soft iron core passes through both coils.

An alternating current passed through the primary coil produces an alternating magnetic field, a field whose strength varies continuously and whose direction reverses periodically. This changing magnetic field causes the magnetic flux linking the secondary coil to change so that an e.m.f. is induced in the secondary coil. The iron core maximizes the magnetic flux linking both coils.

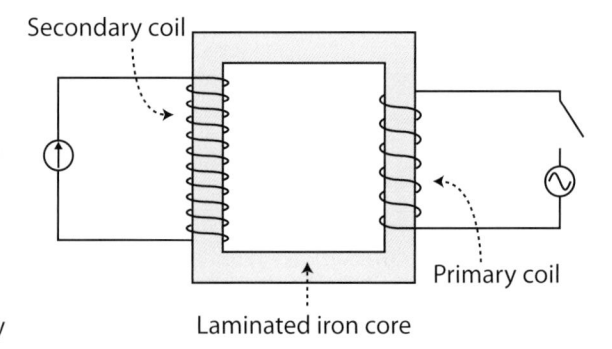

In all transformers the ratio of the turns on each coil determines the ratio of the two voltages:

$\dfrac{N_S}{N_P} = \dfrac{V_S}{V_P}$ where N_P = number of turns on the primary coil
N_S = number of turns on the secondary coil
V_P = voltage applied to the primary coil (input)
V_S = voltage developed across the secondary coil (output)

If we assume that the efficiency of the transformer is 100% then input power = output power. So:

$I_P \times V_P = I_S \times V_S$

Therefore, **in transformers where there are no energy losses:**

$\dfrac{I_P}{I_S} = \dfrac{V_S}{V_P}$ where I_P = current in the primary coil
I_S = current in the secondary coil
V_P = voltage applied to the primary coil (input)
V_S = voltage developed across the secondary coil (output)

Note carefully the different positions of the P and S on each side of this equation.

Power Losses in a Transformer

The efficiency of real transformers is less than 1 (100%), ie not all of the input electrical energy appears as useful output electrical energy. Some of the various ways in which energy is wasted are as follows:
1. Transformers have resistive heat losses in the wires in the coils.
2. Not all of the magnetic flux of the primary passes through or links the secondary coil.

3. Repeatedly magnetising the iron core in one direction and then reversing the direction of magnetisation results in heating of the iron core.
4. The changing magnetic field induces large currents in the iron core. These are called eddy currents and are very large. They result in heating of the core. These eddy currents are reduced by laminating the core.

Transmission of Electricity

Transformers play an important role in the transmission of electricity from where it is generated in power stations to consumers. At the generating end they step the voltage up before it is connected to the transmission cables. At the consumer end, they step the voltage down for use in appliances. The process is illustrated below.

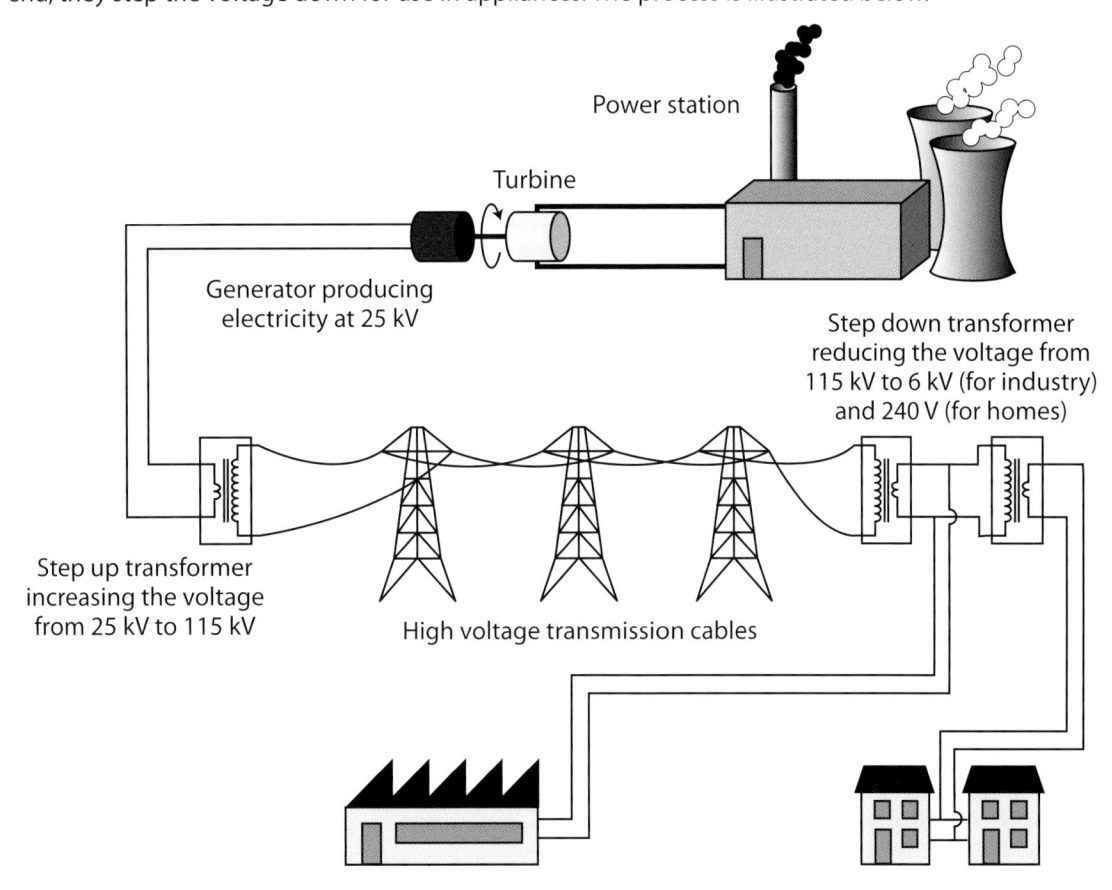

Advantages of High Voltage Electricity Transmission

The cables used to transmit the electrical power from the generator to the consumer have resistance. This means energy is lost as heat due to resistive heating. The diagram on the right is a simplified picture of the electricity generation and transmission system. The power generated is P_{Gen} and the resistance, R, is constant. The power loss in the cables $P_{Loss} = I^2 R$.

This equation suggests that one way to reduce the power loss would be to reduce the resistance, R, of the cables. This could be achieved by using cables of a very large cross section area. However this would considerably increase their weight and hence the cost of construction.

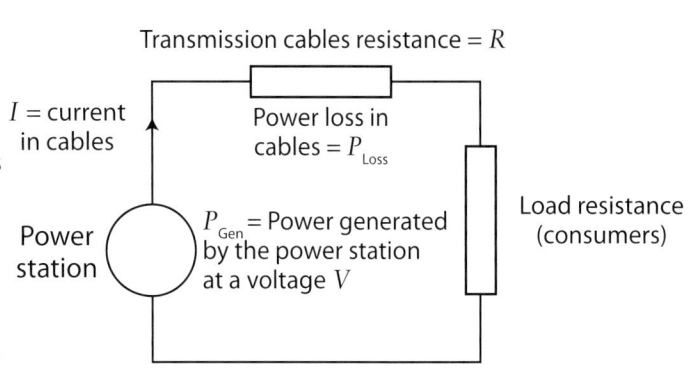

The alternative way to reduce the power loss is to reduce the current, I, using a transformer. This is the function of the step up transformer at the generating station. As the voltage is stepped up, the current is reduced. The electrical power is then transmitted at a high voltage and a low current. What is the impact of reducing the current in this way?

$P_{Gen} = IV$, so therefore $I = \dfrac{P_{Gen}}{V}$ and $P_{Loss} = \dfrac{P_{Gen}^2 R}{V^2}$

Since P_{Gen} and R are both constants this shows that the power loss in the cables P_{Loss} is inversely proportional to the square of the voltage at which the electricity is transmitted to the consumer, ie if the voltage is **doubled** the power loss is reduced by a factor of **four**, and the current I is reduced by a factor of two.

The advantage of this is demonstrated by the following example:

PHYSICS FOR CCEA A2 REVISION GUIDE, 2ND EDITION

Worked Example

A power station generates 400 MW of electrical power at a voltage of 25 kV.
The transmission lines have a resistance of 0.25 Ω per kilometre.
(a) Calculate the energy loss if the power is transmitted at 25 kV.
(b) Calculate the energy loss if a transformer is used to step up the voltage to 115 kV before it is transmitted.

(a) Current $I = \dfrac{P}{V} = \dfrac{400 \times 10^6}{25 \times 10^3} = 1.6 \times 10^4$ A. So power loss $P_{Loss} = I^2 R = (1.6 \times 10^4)^2 \times 0.25 = 6.4 \times 10^7$ W = 64 MW per km

(b) Current $I = \dfrac{P}{V} = \dfrac{400 \times 10^6}{115 \times 10^3} = 3.48 \times 10^3$ A. So power loss $P_{Loss} = I^2 R = (3.48 \times 10^3)^2 \times 0.25 = 3.03 \times 10^6$ W = 3.03 MW per km

Exercise 5.5

1. Identical magnets are set on electronic scales as shown in the diagram. A uniform magnetic field exists between the poles of the magnets. The south pole of each magnet is shaded. The thick current carrying wire between the poles is suspended so that it cannot move. The variable power supply unit is switched off and the balance adjusted to read 0.00 g.

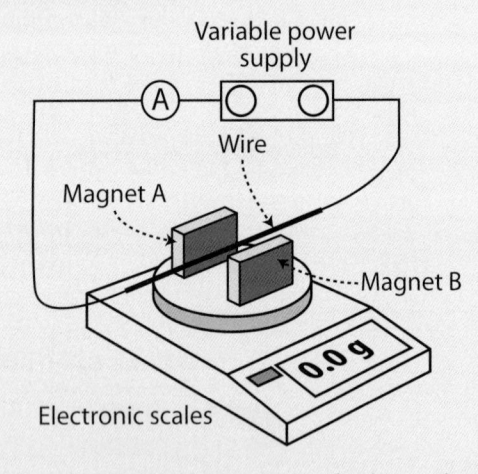

   (a) In what direction is the magnetic field near the thick current carrying wire?

   The variable power supply is now switched on and immediately the reading on the balance rises from 0.00 g to 0.70 g. The reading on the ammeter is 1.40 A.
   (b) Using Newton's third law, state and explain the direction of the force on the wire within the magnetic field.
   (c) Copy the symbol for the variable power supply from the diagram. Indicate on your diagram the polarity of the variable power supply unit by marking a + in one of the circles and a − in the other.
   (d) Calculate the force on the current carrying wire when the variable power supply unit is adjusted to give a current of 2.50 A.
   (e) Calculate the length of wire that is in the magnetic field given that the flux density between the two magnets is 40 mT.

2. (a) State the Laws of Faraday and Lenz in connection with electromagnetic induction.
   (b) Describe how both of these laws might be demonstrated qualitatively using a coil, a bar magnet, a centre zero ammeter and electrical connecting wire.
   (c) A coil of wire, having 10 turns is placed in the magnetic field produced by an electromagnet. The loop of wire has a resistance of 1.2 Ω and an area of 4.0×10^{-3} m². When the electromagnet is switched on it takes 0.8 s to reach its maximum flux density of 120 µT. Calculate the average current that flows in the wire during the 0.8 s after the electromagnet is switched on.

3. The magnetic flux linked with a coil of wire changes with time as shown in the graph below. Copy the graph and on it draw another graph to show how the induced e.m.f. in the coil changes with time.

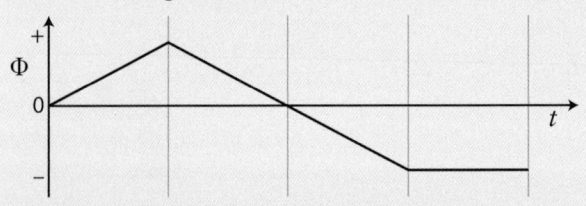

4. Typically, transformers have an efficiency of around 97%.
   (a) State three ways in which transformers lose energy to the environment.
   (b) A step down transformer has a primary coil of 1200 turns and a secondary coil of 30 turns. The primary voltage is 240 V and the primary current is 0.04 A. If the current in the secondary coil is 1.51 A, calculate the transformer's efficiency.

5. The total resistance of the cables connecting a power station to an industrial plant is 0.50 Ω. The power station output voltage is 11 kV and the current supplied to the industrial plant is 100 A.
   (a) Calculate:
       (i) the voltage at the industrial plant,
       (ii) the efficiency of the transmission system.
   (b) Calculate the efficiency of the transmission system if the same power output was transmitted at 1100 V and 1000 A.
   (c) Comment on your answers to parts (a) and (b).

5.6 Deflection of Charged Particles in Electric and Magnetic Fields

Students should be able to:
5.6.1 Demonstrate an understanding that a charge in a uniform electric field experiences a force;
5.6.2 Recall and use the equation $F = qE$ to calculate the magnitude of the force and determine the direction of the force;
5.6.3 Demonstrate an understanding that a moving charge in a uniform, perpendicular magnetic field experiences a force;
5.6.4 Recall and use the equation $F = Bqv$ to calculate the magnitude of the force, and determine the direction of the force.

Motion of Electrons in an Electric Field

Beams of electrons are found in traditional television tubes, cathode ray oscilloscopes, X–ray tubes and electron microscopes. They are produced by an electron gun, as shown in the diagram on the right.

The source of the electrons is a heated wire filament. Electrons with sufficient energy escape from the surface of the filament by a process called **thermionic emission**. The electrons emerge into an electric field created by a large potential difference between the cathode and the anode. This electric field accelerates the electrons, such that they lose electric potential energy and gain kinetic energy. An electron beam emerges through an opening in the anode.

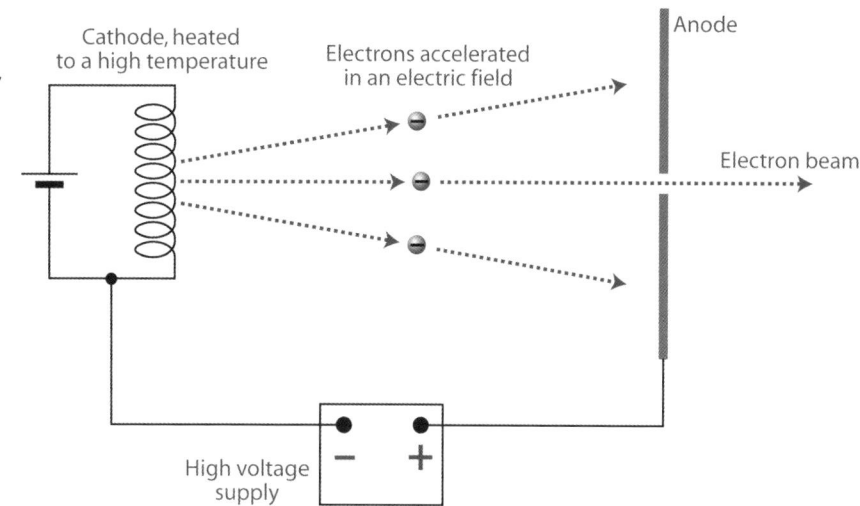

Energy Changes

A volt is defined as a joule per coulomb, which means that a charge of 1 coulomb accelerated through a potential difference of 1 volt will gain 1 joule of energy. Using this definition, and applying the principle of conservation of energy we can say that: **loss of electrical potential energy = gain of kinetic energy**. This can be written in the form of an equation as:

$eV = \frac{1}{2} m_e v^2$ where:
 e = charge on the electron (1.6×10^{-19} C)
 V = potential difference between anode and cathode (V)
 m_e = mass of the electron (kg)
 v = velocity of the electron (m s^{-1})

Other charged particles can be accelerated in a similar way. In this case, the equation above changes to account for the mass and charge of the particle:

$qV = \frac{1}{2} mv^2$ where:
 q = charge on the particle (C)
 V = potential difference between anode and cathode (V)
 m = mass of the particle (kg)
 v = velocity of the particle (m s^{-1})

Force on a Charged Particle

The force on a charged particle in an electric field is given by:

$F = qE$ where:
 F = the force (N)
 q = charge on the particle (C)
 E = electric field strength in V m^{-1} or N C^{-1}

The direction of an electric field is the direction in which a positive charge will experience a force when placed in the field (shown in the left diagram). Remember that electrons have a **negative** charge so they will experience a force in a direction **opposite** to that of the electric field (shown in the right diagram).

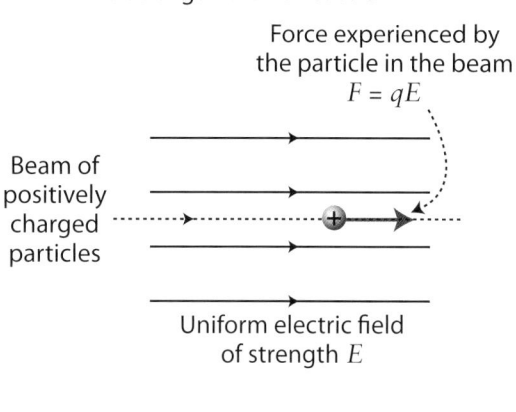

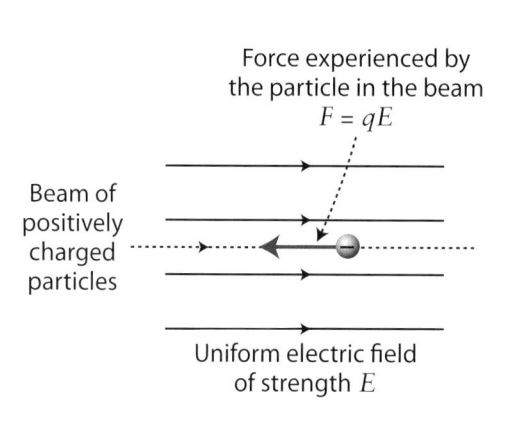

PHYSICS FOR CCEA A2 REVISION GUIDE, 2ND EDITION

Worked Example

Positive ions, each of mass 7.75×10^{-27} kg and charge 3.50×10^{-19} C are accelerated in a vacuum from rest to a speed of 4.25×10^4 m s^{-1}. Calculate the potential difference through which the ions are accelerated to give them this speed.

Loss of electrical potential energy = gain of kinetic energy, so:
$½mv^2 = qV$
$½ \times 7.75\times10^{-27} \times (4.25\times10^4)^2 = 3.50\times10^{-19} \times V$
$7.00\times10^{-18} = 3.50\times10^{-19} \times V$
This gives: $V = 7.00\times10^{-18} \div 3.50\times10^{-19} = 20.0$ V

Deflection of Charged Particles in an Electric Field

The diagram on the right shows what happens when a charged particle is projected into an electric field at right angles to the direction of the field. The charged particle experiences a force in the **vertical direction only**. There is no horizontal force. This means that we treat its motion in the following way:
- Horizontally – constant velocity
- Vertically – uniform acceleration from rest

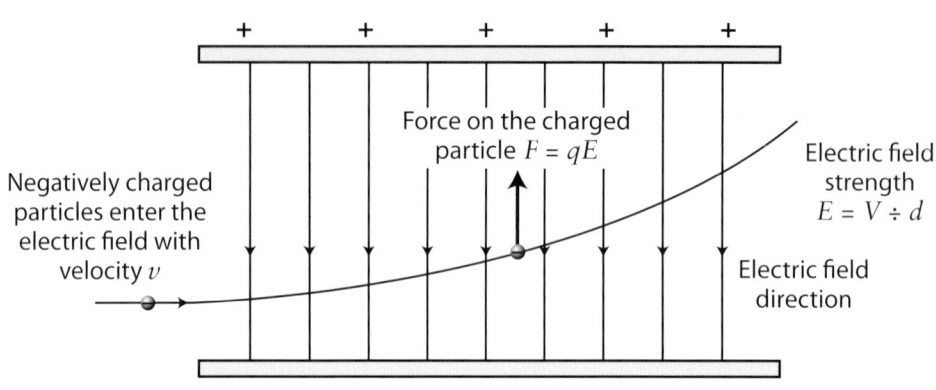

Negatively charged particles enter the electric field with velocity v

Force on the charged particle $F = qE$

Electric field strength $E = V \div d$

Electric field direction

The vertical force is given by $F = qE$. If the separation of the plates is d, then electrical field strength, $E = \dfrac{V}{d}$. So we can write:

$F = qE = q\dfrac{V}{d}$ where: F = the force (N)
q = the charge on the particle (C)
E = electric field strength (V m^{-1} or N C^{-1})
V = potential difference between the plates (V)
d = the separation of the plates (m)

The acceleration of the particle is given by:

$a = \dfrac{qE}{m} = \dfrac{qV}{dm}$ where: a = acceleration of the particle (m s^{-2})
m = mass of the particle (kg)

The equations of motion can be applied to the motion of the charged particle as it moves through the electric field.

At time $t = 0$:   At time $t = T$:
$v_x = v$ $v_x = v$
$v_y = 0$ $v_y = \dfrac{qV}{dm}T$

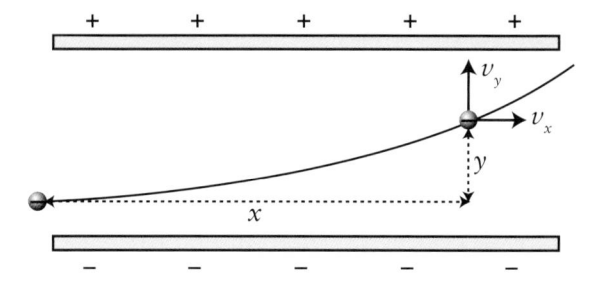

So distances x and y at time $t = T$ are given by:
$x = vT$
$y = 0 + ½aT^2 = ½\dfrac{qV}{dm}T^2$

Worked Example

A plastic sphere of mass 7.82×10^{-14} kg carries a positive charge. It is held stationary between two parallel metal plates with a potential difference between them. This creates an electric field strength of magnitude 6.66×10^4 V m^{-1}.
(a) Draw a diagram of the sphere and show the forces acting on it.
(b) Calculate the charge on the sphere.
(c) Hence calculate the number of elementary charges carried by the sphere.

(a) ↑ Electric force = qE
    ↓ Weight = mg

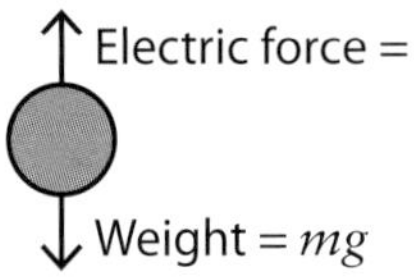

(b) Since the sphere is stationary, we know that $qE = mg$. So:
$q \times 6.66\times10^4 = 7.82\times10^{-14} \times 9.81$ giving: $q = 1.52\times10^{-17}$ C

62

5.6 DEFLECTION OF CHARGED PARTICLES IN ELECTRIC AND MAGNETIC FIELDS

(c) Let N be the number of elementary particles (electrons).
$q = Ne$, so $N = \dfrac{q}{e} = \dfrac{1.52 \times 10^{-17}}{1.6 \times 10^{-19}} = 95$ elementary charges

Deflection of Charged Particles in a Magnetic Field

A moving charge in a magnetic field experiences a force which is perpendicular to both the velocity of the particle and the direction of the magnetic field. Fleming's left hand rule can be used to find the direction of this force, as shown in the diagram on the right.

Note carefully: To determine the direction of the force on an electron moving in a magnetic field you must remember that the movement of the electron is **opposite** to that of positive charge.

Since the force always acts at right angles to the velocity of the charged particles it causes the particles (electrons, protons, ions) to move in circular paths. The diagrams below show the paths taken by beams of negatively charged and positively charged particles when they enter a magnetic field.

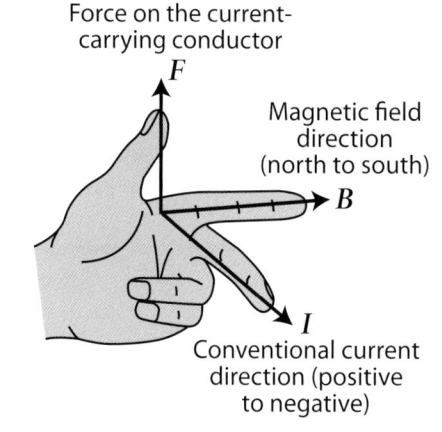

Force on the current-carrying conductor
F
Magnetic field direction (north to south)
B
I
Conventional current direction (positive to negative)

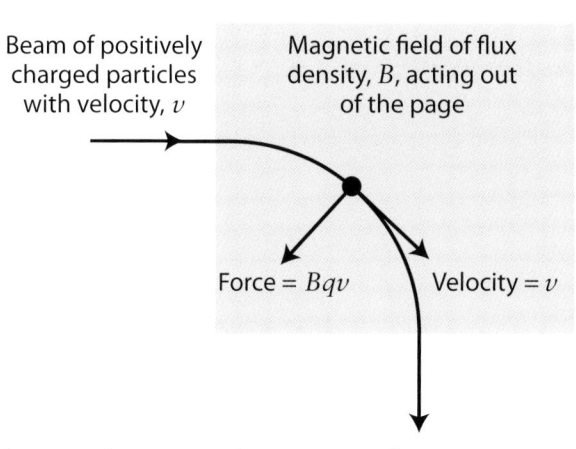

Beam of positively charged particles with velocity, v
Magnetic field of flux density, B, acting out of the page
Force = Bqv Velocity = v

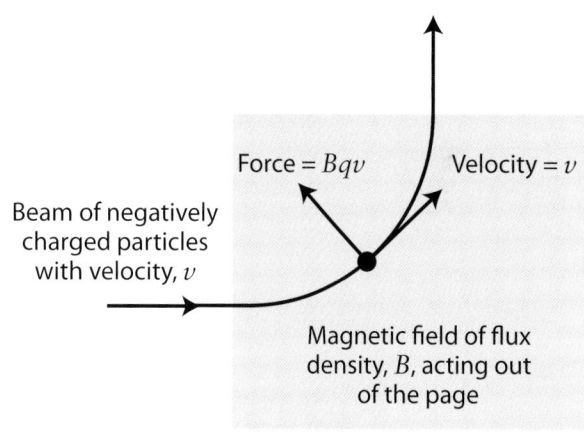

Beam of negatively charged particles with velocity, v
Force = Bqv Velocity = v
Magnetic field of flux density, B, acting out of the page

The force on the moving charge is given by:
$F = Bqv$ where: F = the force (N)
B = magnetic field strength (T)
q = charge on the particle (C)
v = velocity of the particle (m s^{-1})

Since the direction of the force is always at right angles to the plane containing B and v, the charged particles will generally move in a circular arc. **Note:** If the charged particles enter the magnetic field parallel or anti-parallel to the magnetic field lines they do **not** experience any force:

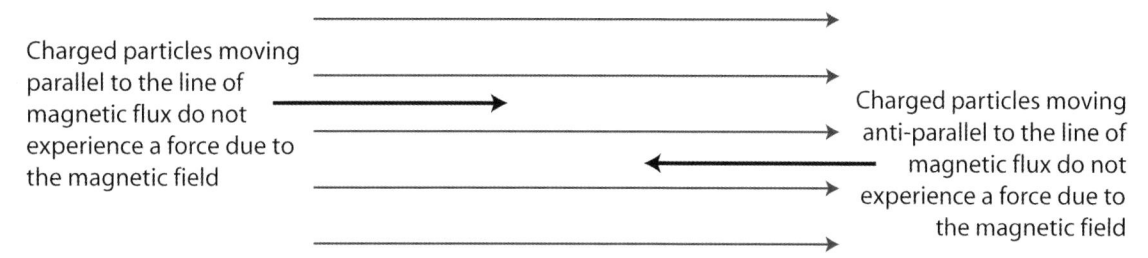

Charged particles moving parallel to the line of magnetic flux do not experience a force due to the magnetic field

Charged particles moving anti-parallel to the line of magnetic flux do not experience a force due to the magnetic field

Exercise 5.6

1. A proton enters the uniform electric field between two horizontal plates, as shown in the diagram below. The region of the electric field is shaded. The proton enters horizontally with a speed $v_0 = 4.00 \times 10^5$ m s^{-1}. The voltage between the plates is 148 V.

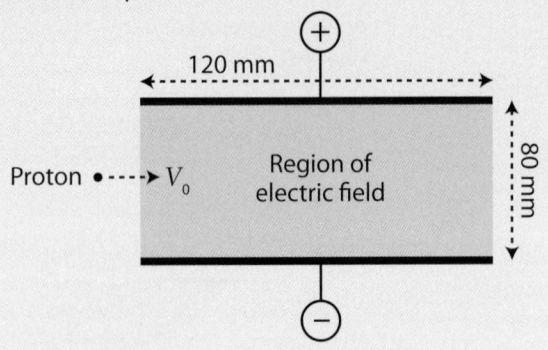

   (a) (i) Calculate the magnitude of the electric field strength E.
   (ii) Calculate the magnitude of the acceleration experienced by the proton if the electric field exerted a constant force of 2.96×10^{-16} N. Ignore the effect of gravity on the proton.
   (b) Calculate the magnitude and direction of the velocity of the proton on exiting the electric field. State the direction relative to the horizontal.

2. An electron beam enters a magnetic field of strength 100 mT with a speed of 8.8×10^7 m s^{-1} in a direction which is perpendicular to the field. The field causes the beam to move in a curved path.
   (a) State the nature of the curved path taken by the electrons.
   (b) If the radius of curvature of their path is 5 mm, find the charge to mass ratio of the electrons.
   (c) What two differences might an experimentalist observe if a proton beam had entered the same field with the same velocity?

5.7 Particle Accelerators

Students should be able to:

5.7.1 Describe the basic principles of operation of a synchrotron;

5.7.2 Demonstrate an understanding of the concept of a relativistic mass increase as particles are accelerated towards the speed of light;

5.7.3 Demonstrate an understanding of the concept of antimatter and that it can be produced using the collisions of high-energy particles from the accelerators;

5.7.4 Describe the process of annihilation in terms of photon emission, and conservation of charge, energy and momentum.

Relativistic Mass Increase

Particle accelerators accelerate subatomic particles to speeds almost equal to the speed of light, and then crash them into one another to see what happens. Einstein showed that energy itself has mass (remember $E = \Delta mc^2$), so that a moving object has a greater mass than an object at rest. The mass of an object that is at rest is known as its **rest mass** or **proper mass**.

When objects move, their mass is observed to increase, what we call the **relativistic mass**. This increase in mass is very, very small when the object is moving at the speeds we experience on Earth, such as the speed of a bus, an aircraft or even a ballistic missile. But when the speed approaches the speed of light, then the observed increase in mass becomes very significant.

Einstein's relativistic mass formula is:

$m = \dfrac{m_o}{\sqrt{1 - \left(\dfrac{v}{c}\right)^2}}$ where m = mass of the object when moving at a speed, v, ie its relativistic mass (kg)
m_o = mass of the object at rest, ie its proper mass (kg)
v = speed of the object (m s^{-1})
c = speed of light (m s^{-1})

The graph on the right illustrates how mass increases with speed. Note that:
- at low speeds the relativistic mass is almost identical to the rest mass; and
- a consequence of the equation is that no object with a rest mass can exceed the speed of light.

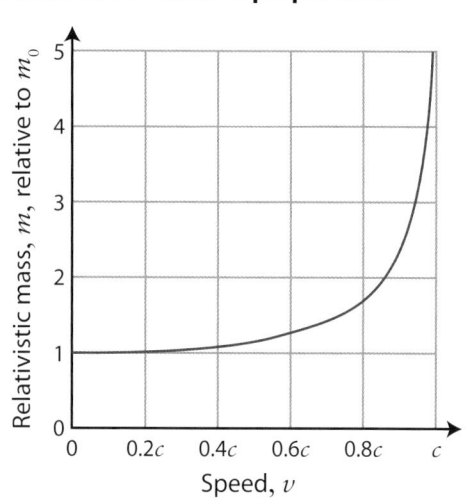

The primary purpose of an accelerator is to **not** to increase the velocity of particles, but to increase the **energy** of the particles. Once a particle is travelling at, say, 99% of the speed of light it is not going to increase its velocity very much, no matter how much more energy is supplied. However, its mass increases as it gains energy. For example, when the speed of a particle increases from $0.99c$ to $0.999c$ the mass increases by a factor of 3. The velocity increase is about 1% but the mass increase is nearer 300%.

The Synchrotron

In the synchrotron, particles are accelerated and held in a circle of fixed radius by means of a magnetic field, the strength of which is varied to ensure that the particles follow a circular path. The acceleration of the particles is achieved by the application of a high frequency alternating voltage located at various cavities along the circumference of the ring.

As the particles accelerate the magnetic field is increased to keep them on the same orbit. As the speed of the particles increases, so too does the magnetic field provided by the powerful magnets. This keeps the charged particles in an orbit of fixed radius. Since the particles take less and less time to complete their orbit so the frequency of the accelerating alternating voltage must increase as well. The magnets perform two functions in the synchrotron – they bend the beam into a circular path and focus it to keep as many particles as possible on the ideal orbit.

Effective focusing of particle beams is very important as it:
- increases the beam intensity, ie more particles per unit area per second.
- reduces the area of cross section of the particle beam, to allow the use of smaller evacuated beam tubes and smaller gaps between magnetic poles.

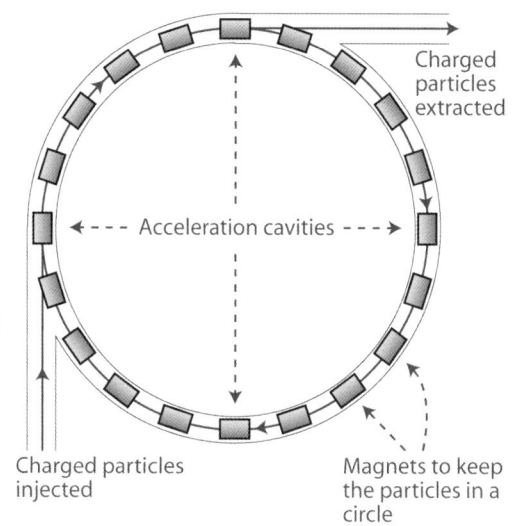

Advantages and Disadvantages of the Synchrotron

A key advantage of a synchroton is that particles can be extracted at various points along the path allowing a number of different experiments to be carried out. A disadvantage is that the accelerated particles lose energy by emitting electromagnetic radiation (called synchrotron radiation) when they move in a circle. The effect can be reduced by increasing the radius of the circle. Synchrotrons are also very expensive to build.

Antimatter

In particle physics, antimatter is a material composed of **antiparticles**. Antiparticles have the same mass as particles of ordinary matter and in some cases opposite electric charge. For example, the antiparticle of the electron is the positively charged antielectron, or positron. The antiparticle of the proton has the same mass as the proton but has a negative charge. In a few cases a particle is its own antiparticle.

Particle–antiparticle pairs can **annihilate** each other (cease to exist) in line with the principle of **conservation of charge**. In their place we observe two gamma ray photons, as shown on the right. The two gamma ray photons are emitted in opposite directions in line with the principle of **conservation of momentum**.

For example, the antielectrons produced in natural radioactivity meet electrons resulting in annihilation and producing pairs of gamma ray photons. When a positron ($^0_1$e) and an electron ($^{\ \ 0}_{-1}$e) meet they annihilate each other, also resulting in two gamma ray photons.

The energy released is obtained from Einstein's mass-energy equation $E = \Delta mc^2$, thus obeying the principle of **conservation of energy**.

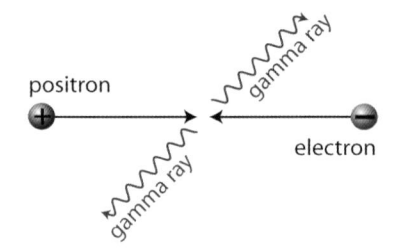

Worked Example

A electron and a positron meet and annihilate each other, producing two gamma ray photons.
(a) What is the energy of each gamma ray photon?
(b)) What is the wavelength of each gamma ray photon?
The mass of an electron is 9.1×10^{-31} kg.

(a) A positron has the same mass as an electron, so the total mass = $2 \times 9.1 \times 10^{-31}$ kg = 18.2×10^{-31} kg.
Total energy produced, $E = \Delta mc^2 = 18.2 \times 10^{-31} \times (3 \times 10^8)^2 = 1.64 \times 10^{-13}$ J (or 1.02 MeV).
Since there are two photons, each one has $1.64 \times 10^{-13} \div 2 = 0.82 \times 10^{-13}$ J (or 0.51 MeV).

(b) Wavelength of each photon = $\dfrac{hc}{E} = \dfrac{6.63 \times 10^{-34} \times 3 \times 10^8}{0.82 \times 10^{-13}} = 2.43 \times 10^{-12}$ m.

Antimatter is composed of antiparticles in the same way that normal matter is composed of particles. This means that an antiproton and a positron can form an antihydrogen atom, which has almost exactly the same properties as a hydrogen atom. Antimatter in very short–lived. Antihydrogen atoms survive for only 40 billionths of a second (4×10^{-11} s) before annihilation with ordinary matter takes place. Ordinary matter is the dominant type of matter in the universe.

Exercise 5.7

1. (a) (i) Draw a large, labelled diagram showing the structure of a synchrotron.
   (ii) Describe briefly the operation of a synchrotron, paying particular attention to the method used to accelerate the particles and how the particles are made to move in a circle of fixed radius.
   (b) (i) When accelerated to very high speed all particles exhibit a relativistic mass change. Explain what this means.
   (ii) What is done in a proton synchrotron to ensure that particles continue to be accelerated, despite the relativistic mass change?
   (c) Protons moving a speed of 2.94×10^8 m s^{-1} in a synchrotron have a relativistic mass of 8.35×10^{-27} kg. Calculate the strength of the local magnetic field required to keep them moving in a circle of radius 10 km.

2. The discovery of the Higgs boson was announced by physicists at CERN in 2012. It has a mass about 126 times that of a proton and has no antiparticle. Estimate the amount of energy needed to make a Higgs boson. Give your answer in GeV.

3. Electron-positron pairs can be observed when cosmic rays interact with matter. Estimate the wavelength of the cosmic ray required and state to what region of the electromagnetic spectrum it belongs.

4. Although never observed, radioactive nuclei of antimatter would emit anti-alpha particles. Describe what an anti-alpha particle would consist of.

5.8 Fundamental Particles

Students should be able to:

5.8.1 Explain the concept of a fundamental particle;

5.8.2 Identify the four fundamental forces and their associated exchange particles;

5.8.3 Classify particles as gauge bosons, leptons and hadrons (mesons and baryons);

5.8.4 State examples of each class of particle;

5.8.5 Describe the structure of hadrons in terms of quarks;

5.8.6 Understand the concept of conservation of charge, lepton number and baryon number;

5.8.7 Describe β-decay in terms of the basic quark model.

Fundamental Particles

An elementary particle, or fundamental particle, is **a particle not known to be made up of smaller particles**. A fundamental particle has no substructure; it is one of the basic building blocks of the universe from which all other particles are made.

Fundamental Forces of Nature

In order of strength, from strongest to weakest, the four fundamental forces are:

- The **strong nuclear force** exists between neutrons and protons in the nucleus. It is clearly strong enough to overcome the electrical repulsion of the protons. It is a **very short range** force and only exists when neutrons and protons are within a distance of around 10^{-15} m of each other. The strong nuclear force determines the structure of the nucleus.
- The **electromagnetic force** affects particles with charge and has an **infinite** range. Electromagnetic forces determine the structure of atoms as well as determining the properties of materials and the results of chemical processes. The electromagnetic force can be **repulsive** or **attractive**.
- The **weak interaction** is the name given to the force that induces beta decay. Beta decay occurs when a neutron decays to a proton and creates an electron and antineutrino in the process. The neutral antineutrino is not affected by the electromagnetic force or the strong nuclear force. The weak interaction is the **short range** force needed to explain this effect.
- **Gravity** affects particles with mass. It is very weak, and is only noticeable when at least one large mass is present. Our weight is due to the gravitational attraction between ourselves and the Earth which has a mass of 6×10^{24} kg. Gravity is always attractive and has an infinite range. Gravity is the force that determines the structure of large scale matter such as stars and galaxies.

Classification of Particles

Particle accelerators allowed physicists to study the nucleus and the interactions of neutrons and protons that form it. Their experiments studied the collisions of high energy particles produced by accelerators and sophisticated detectors surrounding the collision point were used to identify each of the many particles that may be produced in a single collision. Hundreds of different particles have been identified. These new particles have a wide range of properties. One classification of these particles that emerged from all the observations was is the **Standard Model** in which all particles are classified as:

- Gauge bosons
- Leptons
- Hadrons (pronounced haedrons)

Hadrons

These particles are **not** fundamental, that is they are made up of other particles. Hadrons are all affected by the strong interaction, the force that acts between neutrons and protons within the nucleus. it became convenient to divide the hadrons into two sub-groups, baryons and mesons as shown on the right.

The **neutron** and **proton** are both **baryons**. The proton has slightly less mass than the neutron. The neutron and proton are given baryon numbers, B, of 1. Their antiparticles have a baryon number, B, of −1.

Mesons have a mass less than the proton but greater than the electron. The pi–meson family consisting of two charged particles π^+, π^- and the neutral π^0 play a role in the strong nuclear force. The mesons have a baryon number, B, of 0 as do their antiparticles, since they are not baryons. The baryon number, B, is a quantity that is **conserved** during interactions.

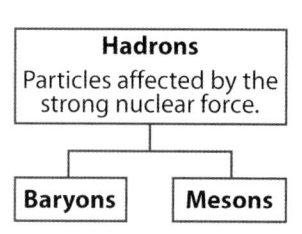

PHYSICS FOR CCEA A2 REVISION GUIDE, 2ND EDITION

Leptons

Leptons are particles that are not affected by the strong interaction. Leptons are fundamental particles, ie they cannot be broken into smaller particles. There are three 'generations' of leptons, the **electron** (e), the **muon** (μ), and the **tau** (τ) particle and their associated neutrinos. Each generation has a greater mass than the one before it. However only the electron and the neutrino occur in normal matter. Each lepton is given a lepton number, L, of 1 and their antiparticles –1. The lepton number, L, is **conserved** during interactions.

Gauge Bosons – Exchange Particles and the Fundamental Dorces

The modern understanding of the four fundamental forces is that they can be treated as the **exchange of particles**. These exchange particles are the **gauge bosons**. Each fundamental force is attributed to the exchange of at least one gauge boson.
- **Photons** are the gauge bosons of the electromagnetic interaction, such as the repulsion between two electrons.
- The **W and Z bosons** are the exchange particles of the weak interaction which governs beta decay.
- **Gluons** play a role in the strong interaction, ie the force that exists between neutrons and protons.
- **Gravitons** are believed to play a similar role in gravity. However the graviton, unlike the other exchange particles, has yet to be detected.

The table below summarises the fundamental forces and their exchange particles.

Force	What it does	Strength (Comparative)	Range	Exchange particle (gauge boson)
Strong nuclear	Holds the nucleus together	1	1×10^{-15} ~ diameter of a nucleus	Gluons
Electromagnetic	Attractive and repulsive force between charged particles	$\sim \frac{1}{150}$	Infinite	Photon
Weak interaction	Induces beta decay	1×10^{-6}	1×10^{-18} m ~ diameter of a proton	W and Z bosons
Gravity	Attractive force between masses	$\sim 1 \times 10^{-39}$	Infinite	Graviton (by analogy only)

The Quark Model of the Hadrons

Baryons (neutrons and protons) and mesons are made up of smaller particles, now known as **quarks**. In the **quark model** the neutron and proton are made up of three quarks while the mesons are made up from just two quarks. The quark model has been confirmed by many observations from particle accelerator experiments. These reveal that there are six types of quarks. A free quark cannot exist: they are always combined in twos (mesons) or in threes (baryons). Antiquarks are the antimatter partners of quarks, they have the same masses as, but the opposite charge from, the corresponding quarks. When a quark meets an antiquark, they may annihilate.

Quarks have fractional electric charges such as $\frac{2}{3}e$. The six types, or 'flavours', of quark are shown in the table on the next page.

5.8 FUNDAMENTAL PARTICLES

Generation	Quark	Symbol	Charge, Q	Baryon number, B
1	up	u	$+\frac{2}{3}e$	$\frac{1}{3}$
1	down	d	$-\frac{1}{3}e$	$\frac{1}{3}$
2	strange	s	$-\frac{1}{3}e$	$\frac{1}{3}$
2	charm	c	$+\frac{2}{3}e$	$\frac{1}{3}$
3	top	t	$+\frac{2}{3}e$	$\frac{1}{3}$
3	bottom	b	$-\frac{1}{3}e$	$\frac{1}{3}$

Antiquark	Symbol	Charge, Q	Baryon number, B
anti-up	$\bar{u}$	$-\frac{2}{3}e$	$-\frac{1}{3}$
anti-down	$\bar{d}$	$+\frac{1}{3}e$	$-\frac{1}{3}$
anti-strange	$\bar{s}$	$+\frac{1}{3}e$	$-\frac{1}{3}$
anti-charm	$\bar{c}$	$-\frac{2}{3}e$	$-\frac{1}{3}$
anti-top	$\bar{t}$	$-\frac{2}{3}e$	$-\frac{1}{3}$
anti-bottom	$\bar{b}$	$-\frac{1}{3}e$	$-\frac{1}{3}$

Note: the CCEA specification only requires knowledge of the up and down quarks and their antiparticles.

Structure of Baryons and Mesons
In the quark model **baryons** (neutrons and protons) consist of three quarks, as shown below:

The **proton** consists of 2 up quarks and 1 down quark.

Quark   u   u   d

Charge $= \frac{2}{3}e + \frac{2}{3}e - \frac{1}{3}e = 1e$

Baryon Number $= \frac{1}{3} + \frac{1}{3} + \frac{1}{3} = 1$

The **neutron** consists of 2 down quarks and 1 up quark.

Quark   u   d   d

Charge $= \frac{2}{3}e + \left(-\frac{1}{3}e\right) + \left(-\frac{1}{3}e\right) = 0$

Baryon Number $= \frac{1}{3} + \frac{1}{3} + \frac{1}{3} = 1$

In the quark model **mesons** consist of two quarks, as shown below.

The **π⁰ meson** consists of 1 up quark and 1 anti–up quark.

Quark   u $\bar{u}$

Charge $= \frac{2}{3}e + \left(-\frac{2}{3}e\right) = 0$

Baryon Number $= \frac{1}{3} + \left(-\frac{1}{3}\right) = 0$

Beta Decay and the Quark Model

The weak interaction force induces beta decay. Inside the nucleus a neutron changes to a proton plus an electron (β⁻) and an antineutrino. This process is given by the equation:

$$^1_0n \rightarrow {}^1_1p + {}^0_{-1}e + {}^0_0\bar{\nu}$$

In terms of the quarks that make up the neutrons, the process involves one of the down quarks that make up the neutron changing to an up quark. This process is given by the equation:

$$^{\frac{1}{3}}_{-\frac{1}{3}}d \rightarrow {}^{\frac{1}{3}}_{\frac{2}{3}}u + {}^0_{-1}e + {}^0_0\bar{\nu}$$

This is illustrated by the diagram on the right. Note that this is a two stage process:

$d \rightarrow u + W^-$ followed by: $W^- \rightarrow e^- + \bar{\nu}_e$

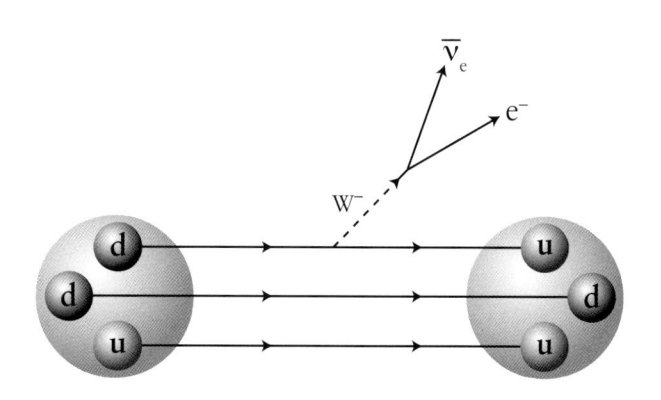

Exercise 5.8

1. (a) State the difference between baryons and mesons in terms of their quark structure.
   (b) State the quark structure for a proton.
   (c) In positron decay a proton decays to a neutron, a positron and an electron-neutrino. Copy and complete the table below with respect to the four particles involved in this decay.

Particle	Charge / C	Baryon number, B	Lepton number, L
Proton			
Neutron			
Positron			
Electron-neutrino			

   (d) Which of the particles identified above are fundamental particles? Give a reason for your answer.

2. The Standard Model asserts that there are only 4 fundamental forces in nature. Copy and complete the table below identifying each force in order of increasing strength, where it might be encountered, its approximate range and associated gauge boson.

Force	Where it is found	Range	Gauge boson
(weakest)			
(strongest)			

Answers

Note: When you have to perform calculations on a set of measurements, the result should be given to the same number of significant figures (sf) as the initial values.

Exercise 4.1

1. A ¼ turn reduces the length of the cube by 0.500 ÷ 4 = 0.125 mm. Hooke's Law constant, $k = F \div x$ = 600 kN ÷ 0.125 mm = 4800 kN mm^{-1} = 4.8×10^6 N m^{-1}

2. (a) Tension in each spring is 36 N
   (b) Extension of first spring = $F \div k$ = 36 ÷ 12 = 3 cm
   Extension of second spring = $F \div k$ = 36 ÷ 18 = 2 cm
   Total extension = 3 + 2 = 5 cm. Hooke's Law constant for combination = $F \div x$ = 36 ÷ 5 = 7.2 N cm^{-1}.

3. (a) See text.
   (b) 1 N causes and extension of 8.5 − 8.0 = 0.5 cm, 4 N causes an extension of 2.0 cm.
   So un-extended length = 8 − 2 = 6 cm.
   (c) Spring constant $k = F \div x$ = 4 ÷ 2 = 2 N cm^{-1}.

4. (a) E = stress ÷ strain = $\sigma \div \varepsilon = (F \div A) \div (\varepsilon \div L) = FL \div A\varepsilon$
   (b) See text.

5. (a)

Load / N	2.45	4.91	7.36	9.81	12.26
Extension /cm	0.8	1.7	2.6	3.4	4.2

   (b) (i) If load and extension are proportional then the ratio should be constant. 2.45 ÷ 0.8 = 3.06; 4.91 ÷ 1.7 = 2.9; 7.36 ÷ 2.6 = 2.8; 9.81 ÷ 3.4 = 2.9; 12.26 ÷ 4.2 = 2.9. Within the limits of experimental error these ratio are within 10% of the mean.
   (ii) k = gradient of the graph with load on the y-axis and extension on the x-axis.

6. (a) Strain ε = extension ÷ original length, so:
   Extension = 8.0×10^{-4} × 1.5 = 1.2×10^{-3} m (1.2 mm).
   (b) Rearranging $E = (FL) \div (Ax)$ gives: $A = (FL) \div (Ex)$ = (8.0 × 1.5) ÷ (1.8×10^{11} × 1.2×10^{-3}) = 5.6×10^{-8} m^2.

Exercise 4.2

1. (a) There must be a fixed mass of gas at constant temperature. So the complete statement should read: "The volume of a **fixed mass** of an ideal gas at **constant temperature** is inversely proportional to the pressure applied to it."
   (b) See text on the Boyle's Law experiment.

2. $\dfrac{p_1 V_1}{T_1} = \dfrac{p_2 V_2}{T_2}$ where p_1 = 270 kPa, T_1 = 12°C = 285 K, T_2 = 25°C = 298 K and $V_1 = V_2$ (so can be ignored). So:
   $\dfrac{270}{285} = \dfrac{p_2}{298}$ giving: $p_2 = \dfrac{270 \times 298}{285}$ = 282.3 kPa

3. $\dfrac{V_1}{T_1} = \dfrac{V_2}{T_2}$ so $\dfrac{50}{293} = \dfrac{V_2}{313}$, giving V_2 = 53.41 cm^3
   The increase in volume is a cylinder.
   Volume of a cylinder = area of cross section × length. So:
   $53.41 = \dfrac{\pi d^2}{4} \times$ length = (3.14 × (5.0×10^{-3})^2 ÷ 4) × length
   So length = distance the mercury moves = 17.4 cm

4. Gradient = $p \div \dfrac{1}{V} = pV$.
   Using $pV = nRT$. This can be rearranged to give:
   $n = pV \div RT$ = gradient ÷ RT
   = (1.5×10^4) ÷ (8.31 × 278) = 6.49 moles

5. (a) N = number of molecules in sample of gas
   m = mass of one molecule
   $<c^2>$ = mean square speed of molecules
   (b) Any 5 of: Point molecules; negligible volume; no intermolecular attractions; the motion of the molecules is random; all collisions are elastic ie kinetic energy and momentum both conserved; the duration of a collision is very much less than the time between collisions.
   (c) $pV = \dfrac{1}{3} Nm <c^2>$
   Nm is the total mass of the gas
   Density ρ = mass ÷ volume = $\dfrac{1}{3}\rho <c^2>$

6. (a) The molar mass of a substance is the mass of one mole of it. One mole of a substance contains Avogadro's number of particles. Avogadro's number is 6.02×10^{23} particles.
   (b) $PV = nRT$, so at constant V and T, the pressure is proportional to the number of moles. Molar masses are: hydrogen = 0.002 kg, helium = 0.004 kg. Number of moles in 5 kg are: hydrogen = 5 ÷ 0.002 = 2500, helium = 5 ÷ 0.004 = 1250. Therefore hydrogen has twice the number of moles compared to helium so the pressure exerted by hydrogen is twice that exerted by helium.

7. $<c^2> = \dfrac{3P}{\rho}$ = (3 × 1.01×10^5) ÷ 1.01 = 300 000 m^2 s^{-2}
   So RMS speed = √300 000 = 547.7 m s^{-1}

8. (a) Fixed mass of gas and constant temperature.
   (b) For diagram see text under Boyle's Law.
   (c) A fixed mass of gas is trapped by oil in the tube. Allow time after each pressure change in order to dissipate any heat generated.

Exercise 4.3

1. Resolving the tension T horizontally and vertically we get:
   $T \sin(90 − \theta) = mg$ and $T \cos(90 − \theta) = mv^2 \div r$. Remember

$\sin(90 - \theta) = \cos \theta = 1.45 \div 1.5 = 0.96$
and $\cos(90 - \theta) = \sin \theta = 0.4 \div 1.5 = 0.27$.
  (a) $T \cos \theta = mg = 0.5 \times 9.81$, $T = (0.5 \times 9.81) \div 0.96 = 5.1$ N
  (b) $T \sin \theta = mv^2 \div r = 5.1 \times 0.27 = 1.4$ N
  (c) Tangential velocity, $mv^2 \div r = 1.4$, giving $v^2 = 1.10$ and therefore $v = 1.05$ m s^{-1}.
  (d) Period of rotation = circumference ÷ tangential velocity = $2\pi \times 0.4 \div 1.05 = 2.39$ s.

2. At the top of the bridge upward reaction force = R and the downward force is the weight mg.
  (a) The resultant of these two forces provides the centripetal force $mv^2 \div r$. So $mg - R = mv^2 \div r$, thus $R = mg + mv^2 \div r = (65 \times 9.81) + (65 \times 152 \div 10) = 637.65 + 1462.5 = 2100.15$ N.
  (b) When $R = 0$, $mg = mv^2 \div r$, so $v^2 = gr = 98.1$ giving $v = 9.9$ m s^{-1}.

3. The string will just be taut when the object is at the top of its motion. The resultant force towards the centre of the circle is $T + mg = mv^2 \div r$.
  (a) $T = 0$, $v^2 = gr = 9.81 \times 1.0$, so $v = 3.13$ m s^{-1}. Time to complete one orbit = $2\pi \times 1 \div 3.13 = 2.00$ s.
  (b) $T = mg + mv^2 \div r$. The maximum tension occurs at the lowest part, so:
  $T_{max} = (0.5 \times 9.81) + (0.5 \times 3.13^2 \div 1.0) = 9.80$ N

4. Max tension in cord is at the bottom of the circle,
  $T = m\omega^2 r + mg$. So: $300 = (1.5 \times \omega^2 \times 1.2) + (1.5 \times 9.81)$
  $\omega^2 = (200 - 1.5 \times 9.81) \div (1.5 \times 1.2) = 102.9$
  $\omega = 10.1$ rad s^{-1}.
  Number of revolutions per second = $10.1 \div 2\pi = 1.6$

5. (a) When an object moves in a circle its velocity is continually changing direction. A changing velocity by definition means an acceleration. This change of velocity is towards the centre of the circle so the acceleration is also directed towards centre of circle.
  (b) The centripetal force is provided by the gravitational attraction of the Earth. It is not a force that appears when objects move in a circle. The gravitational attraction is the unbalanced force that provides the acceleration towards the centre of the Earth. The correct statement should read:

  "The gravitational pull of the Earth provides the centripetal force required to the provide the acceleration towards the centre of the Earth so keeping the spacecraft in orbit."

6. (a) Centripetal force is provided by friction between the turntable and coin. The faster the turntable rotates, the greater the centripetal force required. When the required centripetal force is just greater than the friction force the coin slides off.
  (b) Maximum friction, $F = 0.3mg = m\omega^2 r$, so:
  $\omega^2 = 0.3mg \div mr$. Cancelling m gives: $\omega^2 = 0.3g \div r$
  So: $= 0.3 \times 9.81 \div 0.25 = 11.77$ giving $\omega = 3.43$ rad s^{-1}.

Exercise 4.4

1. (a) Definition of SHM – see text.
  (b) ω^2 is constant measured in s^{-2}. x is the displacement from the fixed point in m. The minus indicates that the acceleration is always in the opposite direction to the displacement.
  (c) In damped oscillations resistive or frictional forces are present which gradually decrease the amplitude of the oscillations. In simple harmonic motion the oscillations are free since no resistive forces act.

2. (a) $A = 6.0\times10^{-2}$ m, $T = 1.05$ s, $\omega = 2\pi \div 1.05 = 5.98$ s^{-1}. So: $a_{max} = \omega^2 A = (5.98)^2 \times 6.0\times10^{-2} = 2.15$ m s^{-2}.
  (b) $x = A \cos \omega t = 6.0\times10^{-2} \times \cos(5.98 \times 2.5)$
  Remember to convert radians to degrees, then:
  $x = 6.0\times10^{-2} \times (-0.731) = -4.39\times10^{-2}$ m, or 4.39 cm below its equilibrium position.
  (c) Frictional or damping forces will gradually transfer the energy of the system into heat so the amplitude of the oscillations will decrease.
  (d) See text – carefully note the difference between the variation with time of these two quantities.

3. (a) l = length of the pendulum in m
  g = acceleration of free fall in m s^{-2}
  M = mass in kg
  k = spring constant in N m^{-1}
  (b) The base unit of T is s. For the simple pendulum the base units of the right-hand side = m^{½} m$^{-½}$ s^1 giving s. For the mass on a spring the base units of the right-hand side = kg^{½} N$^{-½}$ m^{½}. N = kg m s^{-2} so N$^{-½}$ = kg$^{-½}$ m$^{-½}$ s^1
  Collecting terms gives s.

4. (a) Period = time between crests = 13 hours.
  Frequency = $1 \div T = 0.077$ hr^{-1}.
  (b) The water level varies from 1.5 m above the sea bed to 4.8 m above the seabed this is a total distance of 3.3 m so the amplitude of the oscillation is 1.65 m.
  A = amplitude = 1.65 m and $\omega = 2\pi f = 0.48$. So:
  $X = 1.65 \cos(0.48t)$ where t is in hours and A in metres.
  (c) 2.8 m above the sea bed is 0.35 m below the equilibrium position of the motion.
  So: $x = A \cos \omega t$, thus $0.35 = 1.65 \cos(0.48t)$
  $\cos(0.48t) = 0.35 \div 1.65 = 0.21$ (remember this is in radians). $0.48t = 1.36$, giving $t = 2.83$ hours.

5. (a) Damping is caused by resistive frictional forces.
  (b) (i) and (ii) See text
  (c) A forced vibration occurs when a periodic external force makes a system vibrate, for example pushing someone on a spring.
  (d) When the frequency of the driving force is equal to the natural frequency of the vibrating system. This is known as resonance.
  (e) Damping forces resulting in the resonant frequency moving to a lower value (see text).

Exercise 4.5

1. (a) A is the mass number i.e. the sum of neutrons and protons in the nucleus
  (b) 92 protons and 116 neutrons
  (c) $r = r_0 A^{⅓} = 1.2\times10^{-15} \times 208^{⅓} = 7.11\times10^{-15}$ m
  (d) Mass of the nucleus = $92 \times 1.007276 + 116 \times 1.008665$
  = 209.674532 u.
  $209.674532 \times 1.66\times10^{-27}$ kg = 3.4806×10^{-25} kg
  Volume of the nucleus = $\frac{4}{3}\pi r^3 = 1.5058\times10^{-42}$ m^3
  Density = $3.4806\times10^{-25} \div 1.5058\times10^{-42}$
  = 2.311×10^{17} kg m^{-3}.

2. (a) The nucleus has the same charge as the nucleus, ie positive so the α-particles are repelled.
  (b) A small deflection means the nucleus is small and

… ANSWERS

few α-particles get close enough to experience a significant repulsion.
 (c) A back scattered α-particle tells us that the nucleus of the atom is many times more massive than the α-particle. (In the original experiment the gold nucleus was approximately 200 ÷ 4, ie 50 times more massive than the α-particle.)
3. Mass of the oxygen nucleus = $16 \times 1.66 \times 10^{-27}$
 = 2.66×10^{-26} kg.
 Radius of the nucleus $r = 1.2 \times 10^{-15} \times 16^{1/3} = 3.02 \times 10^{-15}$ m
 Volume of the nucleus = $4/3 \pi r^3 = 1.154 \times 10^{-43}$ m^3
 Density = mass ÷ volume = 2.31×10^{17} kg m^{-3}

Exercise 4.6

1. (a) Number of unstable nuclei
 = $(5.0 \times 10^{-3} \div 23.0) \times 6.02 \times 10^{23} = 1.31 \times 10^{20}$
 λ = decay constant = $0.693 \div (2.6 \times 365 \times 86400)$
 = 8.45×10^{-9} s^{-1}
 Activity = $\lambda N = 1.31 \times 10^{20} \times 8.45 \times 10^{-9} = 1.11 \times 10^{12}$ Bq
2. (a) Background activity is subtracted from the measured activity.
 (b) See text, the gradient gives the decay constant and half-life $t_{1/2} = 0.693 \div \lambda$
3. (a) No of ^{12}C atoms in 1 g = $6.02 \times 10^{23} \div 12 = 5.02 \times 10^{22}$
 No of ^{14}C atoms = $1.0 \times 10^{-12} \times$ No of ^{12}C atoms
 = 5.02×10^{10} atoms of ^{14}C.
 (b) Initial activity of ^{14}C = λN
 $\lambda = 0.693 \div (5730 \times 365 \times 86400) = 3.84 \times 10^{-12}$ s^{-1}
 So initial activity = $3.84 \times 10^{-12} \times 5.02 \times 10^{10} = 0.193$ Bq
 (c) $A = A_0 e^{-\lambda t}$, so $0.09 = 0.193\, e^{(-3.84 \times 10^{-12} \times t)}$
 So $0.09 \div 0.193 = e^{(-3.84 \times 10^{-12} \times t)}$
 Taking natural logs: $\ln(0.466) = -3.84 \times 10^{-12} \times t$
 So $t = 1.99 \times 10^{11}$ seconds ≈ 6305 years old.

Exercise 4.7

1. (a) 1 MeV = 1.6×10^{-13} J. The mass-energy equation gives:
 $E = \Delta mc^2$, so $\Delta m = E \div c^2 = 1.6 \times 10^{-13} \div (3 \times 10^8)^2$
 = 1.78×10^{-30} kg. (The rest mass of the electron is 9.1×10^{-28} kg, so this increase in mass is around 0.2%(1/500th) of the rest mass.)
 (b) Energy = current × potential difference × time
 = $13.5 \times 8.0 \times 5 \times 3600 = 1.94 \times 10^6$ J.
 Using $E = \Delta mc^2$ gives $\Delta m = 2.16 \times 10^{-11}$ kg.
2. Total mass of the nucleons
 = $28 \times 1.008665 + 28 \times 1.007276 = 56.446348$ u
 Mass difference (mass defect) = $56.446348 - 55.934938$
 = 0.51141 u = 8.49×10^{-28} kg.
 Total binding energy $E = \Delta mc^2 = 8.49 \times 10^{-28} \times (3 \times 10^8)^2$
 = 7.64×10^{-11} J = 477.5 MeV.
 Binding energy per nucleon = $477.5 \div 56 = 8.53$ MeV.
3. Mass of LHS = $235.04394 + 1.008665 = 236.052605$ u
 Mass of RHS = $139.91728 + 92.92204 + (3 \times 1.008665)$
 = 235.865315 u
 Therefore mass reduction = $236.052605 - 235.865315$
 = 0.18729 u = $0.18729 \times 1.66 \times 10^{-27}$ kg
 Using $E = \Delta mc^2 = 0.18729 \times 1.66 \times 10^{-27} \times (3 \times 10^8)^2$ J
 = $2.7981126 \times 10^{-11}$ J ≈ 2.798×10^{-11} J
4. Mass difference =
 $(3.01605 + 2.01410) - (4.00260 + 1.00866) = 0.001889$ u

Convert to kg:
$0.001889 \times 1.66 \times 10^{-27} = 3.136 \times 10^{-30}$ kg
Using $E = \Delta mc^2 = 2.822 \times 10^{-13}$ J = 17.6 MeV

Exercise 4.8

1. 1 = control rod, 2 = fuel rod, 3 = moderator, 4 = concrete shield. See the text for the role each plays in the reactor.
2. 1. Some neutrons might escape from the fuel rod without causing fission. 2. Some neutrons are absorbed by other nuclei, eg $^{238}_{92}$U, without causing fission. 3. Some cause further fission of $^{235}_{92}$U.
3. Critical size is the minimum amount of fuel need to maintain a chain reaction within the uranium. Below this amount the chain will not be maintained and fission will cease.
4. (a) $^2_1H + ^3_1H \rightarrow ^4_2He + ^1_0n$
 (b) Any two of: 1. This reaction produces a greater energy release than other similar reactions. 2. There is plentiful supply of deuterium from sea water and tritium from lithium. 3. Limited waste products (neutron irradiated materials), hence no long term storage is required.

Exercise 5.2

1. (a) Earth: $T^2/r^3 = (1.00)^2 \div (1.49 \times 10^{11})^3$
 = 3.02×10^{-34} year^2 m^{-3}. Jupiter: $T^2/r^3 = (11.9)^2 \div (7.79 \times 10^{11})^3 = 3.00 \times 10^{-34}$ year^2 m^{-3}. Since T^2/r^3 is (nearly) the same for both planets, Kepler's third law is verified.
 (b) Earth: $T^2/r^3 = 4\pi^2 \div Gm_s$ so: $m_s = 4\pi^2 r^3 \div GT^2$
 = $\{4\pi^2 \times (1.49 \times 10^{11})^3\} \div (6.67 \times 10^{-11} \times (365 \times 24 \times 3600)^2)$
 = 1.96×10^{30} kg
 (c) $F = Gm_s m_j \div r^2$, so $m_j = (Fr^2) \div (Gm_s) = 1.95 \times 10^{27}$ kg
2. Value of g at P due to mass at A = $Gm \div r^2$
 = $6.67 \times 10^{-11} \times 10 \div 52 = 2.67 \times 10^{-11}$ N kg^{-1} towards A. By symmetry g at P due to mass at B also = 2.67×10^{-11} N kg^{-1} towards B. If θ is the angle between PA and the vertical, then θ = sin^{-1}(3 ÷ 5) = 36.87°. Upon resolving, we find that the horizontal components, $2.67 \times 10^{-11} \times \sin θ$, are in opposite directions and cancel out. There are two vertical components, both acting downwards, of $2.67 \times 10^{-11} \times \cos θ$. Therefore the resultant value of g is $2 \times 2.67 \times 10^{-11} \times \cos 36.87 = 4.27 \times 10^{-11}$ N kg^{-1} downwards (along the central vertical line in the diagram).
3. If r is the distance between the centre of Saturn and the geostationary satellite, then: $r^3 = GT^2 m_s \div 4\pi^2$
 = $(6.67 \times 10^{-11} \times (10.8 \times 3600)^2 \times (5.68 \times 10^{26}) \div 4\pi^2$
 = 1.45×10^{24} giving r = 113 200 km. Height above the planet's surface = 113 200 – R_s = 113 200 – 60 475
 = 52 725 km (which is within Saturn's ring system).
4. (a) The field lines should be shown radially inwards with arrows pointing towards the centre of the Earth.
 (b) Gravitational force on a mass m at the surface is given by: $F = mg = Gm_E m \div r^2$ where r is the Earth's radius $m_E = gr^2 \div G = 9.81 \times (6.4 \times 10^6)^2 \div 6.67 \times 10^{-11}$
 = 6.02×10^{24} kg. Mean density, ρ = mass ÷ volume
 = $(gr^2 \div G) \div (4/3 \times \pi r^3) = 3g \div 4\pi Gr$. Mean density,
 ρ = $(3 \times 9.81) \div (4 \times \pi \times 6.67 \times 10^{-11} \times 6.4 \times 10^6)$
 = 5500 kg m^{-3}.
5. Let the distance from B to C be d. At C the magnitude of

73

the field strength due to A is equal to that due to B, so: $(G \times 100) \div (0.5)^2 = (G \times 16) \div d^2$, giving $400 = 16 \div d^2$ so $d^2 = 16 \div 400 = 0.04$, and hence d = 0.2 m. So the distance AB = 0.5 m + 0.2 m = 0.7 m.

Exercise 5.3

1. (a) The electric field strength is the force on a charge of +1 C placed in an electric field.
   (b) (i) and (ii) Note that the field lines are parallel, equally spaced and all pointing towards the negative plate.

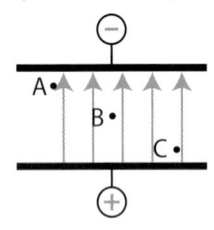

   (iii) The electrical force on each charge is the same in both magnitude and direction. The distances from the upper plate are irrelevant because the field is uniform.
   (iv) $E = V \div d = 75 \div 0.15 = 500$ V m^{-1}.
   (v) $F = Eq = 500 \times 1.6 \times 10^{-3} = 0.8$ N vertically downwards, since the charge is negative.

2. (a) Electrical force = gravitational force, so:
   $$F = \frac{q_1 q_2}{4\pi\varepsilon_0 r^2} = \frac{G m_1 m_2}{r^2}$$
   Since $q_1 = q_2 = Q$ and $m_1 = m_2 = 5000$ kg, and also by multiplying both sides by r^2, we have:
   $8.99 \times 10^9 \times Q^2 = 6.67 \times 10^{-11} \times 5000^2$
   giving $Q^2 = 1.855 \times 10^{-13}$ C^2 and $Q = 4.31 \times 10^{-7}$ C.
   (b) The student is correct. The gravitational/electrical neutral points are both midway between the masses/charges.

3. (a) Weight of ball = 0.001 kg × 9.81 N kg^{-1} = 0.00981 N.
   $\sin \theta = 0.100$, so $\cos \theta = \sqrt{(1 - 0.01)} = \sqrt{0.99} = 0.995$.
   Weight = $T \times \cos \theta = T \times 0.995$, where T = tension in string, giving $T = 0.00981 \div 0.995 = 0.00986$ N.
   (b) Repulsive force = $T \times \sin \theta = 0.00981 \times 0.100$ = 9.81×10^{-4} N.
   (c) Repulsive force = $(1 \div 4\pi\varepsilon_0) \times (q^2 \div r^2)$
   $9.81 \times 10^{-4} = 8.99 \times 10^9 \times q^2 \div (0.20)^2$
   $q^2 = 9.81 \times 10^{-4} \times (0.20)^2 \div 8.99 \times 10^9 = 4.365 \times 10^{-15}$
   $q = \sqrt{(4.365 \times 10^{-15})} = 6.61 \times 10^{-8}$ C.

4. (a) $E = (1 \div 4\pi\varepsilon_0) \times (q \div r^2)$.
   E at C due to +10 mC charge
   = $(8.99 \times 10^9) \times (1 \times 10^{-2})^2 \div 4^2 = 56.19$ kN C^{-1} along BC.
   E at C due to –10 mC charge
   = $(8.99 \times 10^9) \times (1 \times 10^{-2})^2 \div 4^2 = 56.19$ kN C^{-1} along CA.
   The vertical components of E cancel. Each horizontal component = 56.19 × cos 60°. So the total electric field at C = 2 × 56.19 × cos 60° = 56.19 kN C^{-1} horizontally to the left.
   (b) If both charges were +10 mC, the horizontal components would cancel out. The resultant field would be 2 × 56.19 × sin 60° = 102.47 kN C^{-1} vertically upwards.

5. Similar: Both have an infinite range. Both obey an inverse square law with distance. Different: Electrical fields can be shielded (with an insulator), gravitational fields cannot be shielded. The gravitational force is always attractive but the electrical force can be attractive or repulsive.

Exercise 5.4

1. (a) $C = Q \div V = 3 \times 10^{-6} \div 120 \times 10^{-6} = 2.5 \times 10^{-11}$ F
   (b) Energy, $W = \tfrac{1}{2} QV = \tfrac{1}{2} \times 3 \times 10^{-6} \times 120 \times 10^{-6} = 0.18$ J

2. (a) Three 3 µF capacitors in series.
   (b) An arrangement of three 3 µF capacitors in series in parallel with another arrangement of three 3 µF capacitors in series.
   (c) An arrangement of three 3 µF capacitors in series, in parallel with a single 3 µF capacitor.
   (d) An arrangement of three 3 µF capacitors in series, in parallel with another arrangement of three 3 µF capacitors in series and a with a single 3 µF capacitor.

3. (a) (i) $Q = CV = 100 \times 10^{-6} \times 12 = 1.2$ mC.
   (ii) $W = \tfrac{1}{2} CV^2 = 0.5 \times 100 \times 10^{-6} \times 12^2 = 7.2$ mJ.
   (b) (i) Total capacitance of combination
   = 100 µF + 300 µF = 400 µF
   Total charge stored = 1.2 mC, as before:
   $V = Q \div C = 1.2 \times 10^{-6} \div 400 \times 10^{-6} = 3$ V.
   (ii) $W = \tfrac{1}{2} CV^2 = 0.5 \times 400 \times 10^{-6} \times 3^2 = 1.8$ mJ.
   (iii) When current flowed from the 100 µF capacitor to charge the 300 µF capacitor heat was produced in the connecting cable. This heat loss explains why the electrical energy stored in the capacitors at the end is less than that stored at the beginning.

4. (a) (i) $\tau = RC = 50 \times 10^3 \times 100 \times 10^{-6} = 5$ s.
   (ii) $Q = V \times C = 12 \times 100 \times 10^{-6} = 1.2$ mC.
   (iii) The charge approaches a maximum of 1.2 mC:

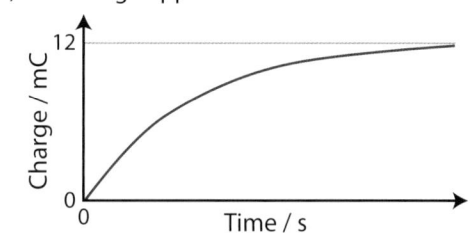

   (iv) As charge accumulates on the capacitor the voltage across the plates increases. When the voltage reaches 12 V the battery's e.m.f. is no longer sufficient to push negative electrons onto the negative plate of the capacitor against the opposing potential difference.
   (v) $Q = Q_0 (1 - e^{\frac{-t}{\tau}})$, and in this case $Q = Q_0 \div 2$, so:
   $0.5 = 1 - e^{\frac{-t}{\tau}}$. We know that $\tau = 5$, so we have:
   $-0.5 = -e^{\frac{-t}{5}}$. Therefore: $0.5 = e^{\frac{-t}{5}}$.
   Taking logs of both sides and rearranging gives:
   $-5 \ln 0.5 = t$, and hence $t = 3.47$ s.
   (b) (i) $I_0 = V_0 \div R = 12 \div 50\,000 = 0.24$ mA.
   (ii) $Q = Q_0 e^{\frac{-t}{\tau}} = 1.2 \times 10^{-3} e^{\frac{-4}{5}} = 5.39 \times 10^{-4}$ C
   (iii) $V = V_0 e^{\frac{-t}{\tau}}$, so $\ln 3 = \ln 12 - (t \div 5)$. This gives:
   $t = 5 \times (\ln 12 - \ln 3) = 5 \times 1.3863 = 6.931$ s.

Exercise 5.5

1. (a) Right to left (ie, from magnet B to magnet A).
   (b) There is a force on the wire due to the presence of a catapult field. Since the scales read a higher value, the

wire must exert a downwards force on the magnet assembly. By Newton's third law, there must therefore be an upwards force on the wire due to the interaction of the field and the current.
(c) By Fleming's left hand rule, the current must flow out of the left terminal of the power supply, so it should be marked +. The right terminal is therefore marked –.
(d) By proportion, the reading on the scales will rise from 0.70 g to (0.70 × 2.50 ÷ 1.40) = 1.25 g, so the force is $1.25 \times 10^{-3} \times 9.81 = 12.3$ mN.
(e) $l = F \div BI = (12.3 \times 10^{-3}) \div (40 \times 10^{-3} \times 2.50)$ m = 12.3 cm

2. (a) Faraday's Law of Electromagnetic Induction states: The magnitude of the induced e.m.f. is equal to the rate of change of magnetic flux linkage. Lenz's Law of Electromagnetic Induction states: The direction of the induced current is such that it opposes the change in the magnetic flux that is producing it.
(b) Set up the apparatus as shown in the diagram:

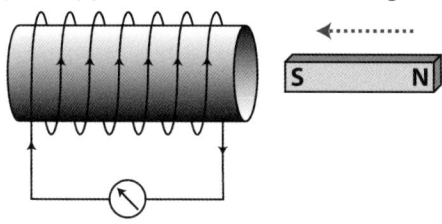

Plunge the south pole of the magnet into the coil and observe that the ammeter needle flicks to the left while the magnet is in motion. This occurs because the magnet causes the magnetic flux linked with the coil to change. Observe that when the magnet is stationary, no current is induced because there is no change in the flux linked with the coil; so the needle on the ammeter returns to zero. Carry out the same procedure again using (i) a stronger magnet and then (ii) plunging the magnet into the coil faster. Both cause the rate of flux linkage to be greater and hence there is a larger induced current. These observations confirm Faraday's Law.

Observe that when the south pole is plunged into the magnet, the ammeter needle flicks to the left showing that the current in the coil is in the direction indicated by the arrows in the diagram. By the right hand grip rule the right side of the coil is therefore a south pole which opposes the incoming magnet, confirming Lenz's Law. If the north pole of the magnet was plunged into the right side of the coil the ammeter needle would flick to the right, indicating that the current in the coil was in the direction opposite to that shown in the diagram, so the right end of the coil was a North pole, again confirming Lenz's Law.
(c) Flux linkage $N\Phi = BAN = 120 \times 10^{-6} \times 4.0 \times 10^{-3} \times 10$ = 4.8×10^{-6} Wb. Rate of change of flux linkage = $N\Phi \div t = 4.8 \times 10^{-6} \div 0.8 = 6 \times 10^{-6}$ V. Average current = $V \div R = 6 \times 10^{-6} \div 1.2 = 5 \times 10^{-6}$ A.

3. In the first period there is a constant positive slope, so the induced e.m.f is constant and negative. In the second and third periods there is a constant negative slope, so the induced e.m.f is constant and positive. In the fourth period the flux is constant, so the induced e.m.f is zero. Your graph should look as follows:

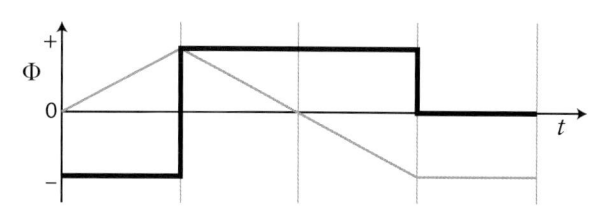

4. (a) (i) Transformers have resistive heat losses due to the wires in the coils. (ii) Not all of the magnetic flux of the primary passes through or links the secondary coil. (iii) The changing magnetic field induces large currents in the iron core. These are called eddy currents and are very large. They result in heating of the core.
(b) Power in primary = $V_p \times I_p = 240 \times 0.04 = 9.6$ W. Secondary voltage = $240 \times 30 \div 1200 = 6$ V. Power in secondary = $V_s \times I_s = 6 \times 1.55 = 9.3$ W. Efficiency = useful power output ÷ total power input = 9.3 ÷ 9.6 = 0.97 = 97%.

5. (a) (i) Voltage lost in cable = $IR = 100 \times 0.50 = 50$ W.
(ii) Voltage supplied = 11 000 – 50 = 10 950 V. Power supplied to industrial plant = VI = 10 950 × 100 = 1095 kW. Efficiency = useful Power out ÷ total Power supplied = 1095 ÷ 1100 = 99.5%.
(b) Power lost in cable = $I^2R = 1000^2 \times 0.5 = 500$ kW. Efficiency = useful power out ÷ total power supplied = (1100 – 500) ÷ 1100 = 54.5%.
(c) Much more power is lost when it is transmitted at a lower voltage.

Exercise 5.6

1. (a) (i) $E = V \div d = 148 \div 80 \times 10^{-3} = 1.85 \times 10^3$ V m^{-1}.
   (ii) $a = F \div m = 2.96 \times 10^{-16} \div 1.66 \times 10^{-27}$ = 1.77×10^{11} m s^{-2}.
(b) Horizontal velocity v_0 remains constant = 4.00×10^5 m s^{-1}. Time spent in the electric field = horizontal distance (120 mm) divided by the constant horizontal velocity = $120 \times 10^{-3} \div 4.00 \times 10^5$ = 3.0×10^{-7} s. Vertical velocity = $u + at$ = $0 + 1.77 \times 10^{11} \times 3.0 \times 10^{-7} = 5.31 \times 10^4$ m s^{-1}. The velocity on exiting the electric field is the resultant of these velocities. Using Pythagoras' Theorem: $v^2 = (4.00 \times 10^5)^2 + (5.31 \times 10^4)^2$ giving $v = 4.04 \times 10^5$ m s^{-1}. The angle θ to the horizontal is obtained by: tan θ = $5.31 \times 10^4 \div 4.00 \times 10^5$, giving θ = 7.56°. This is below the horizontal since the protons are attracted downwards towards the negatively charged plate.

2. (a) A circular path.
(b) Equating magnetic and centripetal force, $Bqv = mv^2 \div r$. So, $q/m = v \div Br$ = $8.8 \times 10^7 \div (0.1 \times 5 \times 10^{-3}) = 1.76 \times 10^{11}$ C kg^{-1}.
(c) The radius of curvature would be about 1840 times greater than that for the electron (since the proton's mass is approximately 1840 times greater than that of the electron). But since that is around 9 metres, the proton would probably move in a circular arc before leaving the field. Further, since the proton and electron have opposite charges, the proton would move in the opposite direction. So if the electron was moving clockwise, the proton would move anti-clockwise and vice versa.

Exercise 5.7

1. (a) (i) See text for diagram of synchroton.
   (ii) A synchrotron is a type of particle accelerator in which the kinetic energy of a charged particle is progressively increased as the particle moves in a highly evacuated tube around a circular track. The particles are accelerated between metal tubes whose electrical potential changes in such a way as to increase the particles' speed as they take energy from the electric field between them. As the particles' speed increases so must the centripetal force to enable them to move at constant radius. This is done by increasing the strength of the electromagnets in the ring.
   (b) (i) Relativistic mass change is the increase in the particles' mass, in a way predicted by Einstein's Special Theory of Relativity, as their speed approaches the speed of light.
   (ii) The potential on the accelerating cavities changes at a variable frequency. As the particles' velocity approaches that of light, the frequency does not increase as rapidly as before, to maintain synchronicity with the particles' speed.
   (c) $B = mv \div rq$
   $= (8.35\times10^{-27} \times 2.94\times10^{8}) \div (10\,000 \times 1.60\times10^{-19})$
   $= 1.53\times10^{-3}$ T.

2. $E = \Delta mc^2 = (126 \times 1.67\times10^{-27}) \times (3\times10^8)^2$
   $= 1.9088\times10^{-8}$ J = 119 GeV.

3. $E = \Delta mc^2 = (2 \times 9.11\times10^{-31}) \times (3\times10^8)^2 = 1.64\times10^{-13}$ J.
   $\lambda = hc \div E = 6.63\times10^{-34} \times 3\times10^8 \div 1.64\times10^{-13}$
   $= 1.21\times10^{-12}$ m. This wavelength is in the gamma-ray region of the electromagnetic spectrum.

4. An anti-nucleus consisting of two anti-protons and two anti-neutrons.

Exercise 5.8

1. (a) A meson consists of a quark and an antiquark. A baryon consists of three quarks.
   (b) Up, up and down (uud).
   (c)

Particle	Charge / C	Baryon number, B	Lepton number, L
Proton	$+1.6\times10^{-19}$	1	0
Neutron	0	1	0
Positron	$+1.6\times10^{-19}$	0	-1
Electron-neutrino	0	0	1

   (d) The positron and electron-neutrino are fundamental. The proton and neutron are not fundamental. They are hadrons, and hadrons are composed of smaller particles. Fundamental particles have no structure, they are not composed of smaller particles.

2.

Force	Where it is found	Range	Gauge boson
Gravity (weakest)	Attractive force between masses	Infinite	Graviton
Weak	In beta decay	$\approx 1\times10^{-18}$ m (1 proton diameter)	W and Z bosons
Electro-magnetic	Force between charged particles	Infinite	Photon
Strong (strongest)	Holds nucleus together	$\approx 1\times10^{-15}$ m (1 nuclear diameter)	Gluon